Animal Rights

Animal Rights

A Subject Guide, Bibliography, and Internet Companion

John M. Kistler

Foreword by Marc Bekoff

Greenwood Press
Westport, Connecticut • London

Library of Congress Cataloging-in-Publication Data

Kistler, John M., 1967–
Animal rights ; a subject guide, bibliography, and Internet companion / John M. Kistler ; foreword by Marc Bekoff.
p. cm.
Includes bibliographical references and index.
ISBN 0-313-31231-1 (alk. paper)
1. Animal rights—Bibliography. 2. Animal rights—Indexes. 3. Animal welfare—Bibliography. 4. Animal welfare—Indexes. I. Title.
Z7164.C45 K57 2000
[HV4708]
016.179′3—dc21 99-088482

British Library Cataloguing in Publication Data is available.

Library of Congress Catalog Card Number: 99-088482
ISBN: 0-313-31231-1

First published in 2000

Greenwood Press, 88 Post Road West, Westport, CT 06881
An imprint of Greenwood Publishing Group, Inc.
www.greenwood.com

Printed in the United States of America

The paper used in this book complies with the Permanent Paper Standard issued by the National Information Standards Organization (Z39.48–1984).

10 9 8 7 6 5 4 3 2 1

Contents

Foreword

ANIMALS AND HUMANS

It is a pleasure to write the Foreword for John Kistler's timely and comprehensive bibliography of animal rights. It is an invaluable resource.

As we head into the twenty-first century, interest in relationships between human beings and animal beings (hereafter "animals") is growing exponentially. In current discussions about the moral status of animals, there is an obvious "progressive" trend for greater protection for wild and captive animals. Time is not on our side.

Many of the numerous issues that need to be considered in the area of animal protection are extremely complex. Indeed, often it seems as if there are inconsistencies among those who truly want to protect animals. As Lisa Mighetto notes in her book *Wild Animals and American Environmental Ethics* (Tucson: University of Arizona Press, 1991, page 121): "Those who complain of the 'inconsistencies' of animal lovers understand neither the complexity of attitudes nor how rapidly they have developed." Even with our inconsistencies and contradictions, we have come a long way in dealing with many, but not all, of the problems. But there is still much work to be done.

Globally, populations of humans are growing rapidly, and many populations of wild animals continue losing their battle with humans. Global biodiversity is rapidly, and perhaps irreversibly, dwindling. In Kenya, for example, 58 percent of the animals in the Tsavo region, about 106,000 large mammals, vanished between 1973 and 1993. Problems such as Kenya's concerning farming, tourism, human interests and needs, and the fate of wild animals are issues worldwide. They demand close attention now because of the enormous uncontrolled growth in the number of humans all over the planet, the decline of habitat where animals can live (in Kenya it is estimated that wild lands are disappearing at a rate of 2 percent a year), and the rampant use of animals to meet human needs and desires.

The numbers themselves tell the grim story. For example, in the United States alone, upwards of 93 million pigs; 37 million cattle; two million calves; six million horses, goats, and sheep; and nearly 10 billion chickens and turkeys are slaughtered for food each year. Numerous animals also are used in experimental research to benefit humans. In 1996, about 1.3 million animals, including 52,000 primates, 82,000 dogs, 26,000 cats, 246,000 hamsters, and 339,000 rabbits were used in the United States. This staggering number does not include rats, mice, and birds who are not legally protected. It is estimated that more than 70 million animals are used annually. One animal dies every three seconds in U.S. laboratories. Members of about 170 species, including at least 10 million vertebrates, are used annually for education in the United States. It has been estimated that about 90 percent of animals used for dissection, including frogs, turtles, and fish, are caught from the wild. In addition to being used for food, research, and education, millions of animals, including dogs, cats, rats, mice, guinea pigs, and rabbits, are used for testing nonessential cosmetic products such as deodorants, shampoos, soaps, and eye make-up. Neither the United States Food and Drug Administration nor the Consumer Product Safety Commission requires that animals be used to test the safety of products, but many companies still use animals for such purposes.

We also need to be deeply concerned about the well-being of our companion ("pet") animals. Michael Tobias, in his book *Voices from the Underground: For the Love of Animals*, (Pasadena, CA: New Paradigm Books, 1999) reports that, according to two surveys taken in 1994, there were about 235 million companion animals in the United States, including 60 million cats, 57 million dogs, 12.3 million rabbits, guinea pigs, hamsters, gerbils, and hedgehogs, 12 million fish tanks (no estimate of numbers of fish), 8 million birds, 7.3 million reptiles, and 7 million ferrets. About $17 billion per year is spent on supplies. Although it is wonderful to think that the sheer number of animals indicates that people truly care about their companions, this is not so. Far too many animals breed, and there are numerous unwanted individuals. Many are ignored or abused when they become burdensome to their human companions. Many are also tortured "for fun." Numerous organizations, including local Humane Societies, have programs directly concerned with the well-being and fate of companion animals. The Humane Society of the United States has a program called First Strike devoted to looking at the relationship between cruelty to animals and cruelty to humans.

It is important to note that convenience and tradition often drive animal use, but neither can adequately defend it, even in biomedical and toxicological research. Unfortunately, the use of animal models often creates false hopes for humans in need. It is estimated that only 1 percent to 3.5 percent of the decline in the rate of human mortality since 1900 has stemmed from animal research. The *New England Journal of Medicine* has called the war on cancer a qualified failure. Over 100,000 people die annually from side effects of animal-tested drugs. Early animal models of polio actually impeded progress on finding a vaccine. As pointed out by the Medical Research Modernization Committee, Dr. Simon Flexner's monkey model of polio misled other researchers concerning the mechanism of infection. He concluded that polio only infected the nervous systems of monkeys. However, research using human tissue culture showed that poliovirus could be cultivated on tissue that was not from the nervous system. Also, chimpanzees have been used to study AIDS, but they do not contract AIDS. Numerous humans die using information from various biomedical models because the diseases from which the animals suffered had to be artificially induced and the course of the disease is different from naturally occurring conditions.

There are also problems with zoos. There is much disagreement about whether zoos should exist. A 1995 Roper poll showed that 69 percent of Americans are concerned about the well-being of animals in zoos, wildlife theme parks, and aquariums. Surely, existing zoos are not going to close tomorrow, and the animals who are kept in cages are not going to be killed or released into the wild. This simply is not possible and it would be unethical to kill the animals or to release animals who might not know how to live outside of cages without human assistance. Unwanted and "surplus" animals also need to be dealt with. Often they are sold, traded, or housed in cages away from the main exhibits.

Another area that is receiving much attention concerns the use of nonanimal alternatives for research and education. Many nonanimal alternatives are readily available, and in numerous instances they are equally or more effective. Jonathan Balcombe, associate director for education at the Humane Society of the United States, has recently summarized some of the studies in which the effectiveness of non-animal alternatives was compared with the effectiveness of using animals. He found that for undergraduates, veterinary students, and medical students, equal knowledge or equivalent surgical skills were acquired using alternatives. Nonanimal models were not less educationally effective (see http://www.hsus.org/programs/research/compare.html). For example, in a study of 2,913 first-year biology undergraduates, the examination results of 308 students who studied model rats were the same as those of 2,605 students who dissected rats. When the surgical skills of 36 third-year veterinary students who trained on soft-tissue organ models were compared to the surgical skills of students who trained on dogs and cats, the performance of each group was the same. Virtual surgery has been shown to be an effective alternative. Also, in a study of 110 medical students, students rated computer demonstrations higher for learning about cardiovascular physiology than demonstrations using dogs. There are numerous ways to find out about alternatives (see www.aavs.org and the Animals in Education fact-sheet published by the American Anti-Vivisection Society [AAVS]; see also www.mindlab.msu.edu; www.enviroweb.org/avar/; and www.pcrm.org/issues/

Animal_Experimentation_Issues/college_alternatives.html).

A very useful source of alternatives can also be obtained from the database NORINA (A Norwegian Inventory of Audiovisuals: oslovet.veths.no/NORINA). Since 1991 information has been collected on audiovisual aids that may be used as animal alternatives or supplements in the biomedical sciences at all levels from primary schools to University. NORINA contains information on over 3,500 audiovisuals and their suppliers. Also, there are centers for the development of alternatives to animals at Johns Hopkins University and the University of California at Davis, and they publish newsletters and other material concerning different alternatives. Animalearn can be reached at 1-800-SAY-AAVS.

There is much that people can do to help other animals. There are always alternatives to cruelty. Jane Goodall developed her Roots & Shoots program that requires groups of young people to participate in projects that benefit the environment, animals, and the human community in their area. The program has two major messages: (a) that every individual matters and can make a difference, every day; and (b) "Only when we understand can we care; only when we care shall we help; only if we help will all be saved" (see_www.janegoodall.org/; and www.janegoodall.org/rs/rs_history.html). One can also contact the pain and distress campaign at the Humane Society of the United States (Martin Stephens at <martinls@ix.netcom.com>).

SCIENCE AND ETHICS

It is in the best traditions of science to ask questions about ethics; it is not anti-science to question what we do when we study other animals. Ethics can enrich our views of other animals, and help us to see that variations among animals are worthy of respect, admiration, and appreciation. The study of ethics can also broaden the range of possible ways in which we interact with other animals without compromising their lives. Ethical discussion can help us to see alternatives to past actions that have disrespected other animals and, in the end, have not served us or other animals well. In this way, the study of ethics is enriching to other animals and to ourselves. If we believe that ethical considerations are stifling and create unnecessary hurdles over which we just jump in order to get done what we want to get done, then we will lose rich opportunities to learn more about other animals and also ourselves. The application of ethical enrichment is a two-way street. Our greatest discoveries come when our relationship with other animals is respectful and not exploitive. In the end, the separation of "us" (humans) from "them" (animals) presents a false dichotomy, the result of which is a distancing that erodes rather than enriches the possible numerous and intimate relationships that can develop among all animal life.

SEARCHING THE LITERATURE ON ANIMAL RIGHTS

Much of the interdisciplinary literature dealing with animal rights — the moral obligations that we have to animals — is scattered widely in numerous publications (books, professional journals, popular magazines, pamphlets, and reports). Many are difficult to find. As I discovered when I edited my *Encyclopedia of Animal Rights and Animal Welfare*, numerous people representing many disciplines are interested in animal protection. Interested parties include biologists, philosophers, psychologists, sociologists, anthropologists, theologians, historians, veterinarians, lawyers, educators, public health officials, and ethnologists. It simply is impossible to keep track of their contributions.

John Kistler's bibliography makes this information readily accessible. It is the longest annotated bibliography on animal rights and it has remarkable breadth. Entries deal with general issues in animal rights, uses of animals that result in their death and those that do not, information on endangered species, feral animals, animal intelligence, consciousness, sentience, emotions, and religion. Most entries are from 1985 to the present, with the exception of some seminal historical works. The inclusion of historical references is important for there is a long, rich, and diverse history of events that have centered on human-animal interactions. Furthermore, for each book, the ISBN number is given, which makes it easy to order from bookstores or from the World Wide Web. There are numerous quotations in the annotations and links to authors' websites and reviews where possible. Because this bibliography is sorted by subject, related entries are easy to locate. Even if local libraries do not carry a particular item, this information is easily accessible. Furthermore, there are more than 1,200 Internet links to authors, articles, biographies, and pertinent issues, and this alone makes the volume incredibly valuable.

I am delighted that John Kistler has taken the time to compile this bibliography. He has been working on it for many years. It is an invaluable source of information, the most comprehensive and user-friendly bibliography ever compiled on the subject of animal protection. I hope that all school and public libraries will carry it.

It is important to remember that we are the voice for all voiceless animals and issues concerned with their protection demand close inspection and open discussion. All animals deserve our admiration, compassion, and respect. Certainly we cannot continue to treat animals the way we have treated them in the past. We, and all animals with whom we share our time and space, should be viewed as friends and partners in a joint venture. I hope that this collection will go a long way toward fostering these relationships.

Marc Bekoff
Professor of Biology
University of Colorado at Boulder

Acknowledgments

I wish to thank my family for inspiring me to a special love for animals, and Rachel George for interesting me in a career in library science.

I appreciate the support of many members of my church, for being there during difficult times, especially the Monsmas, the Willsons, N. Kunsak, and K. Osterhout. I will also remember the teaching of Dr. John Gerstner and his wife Edna; the Christian world misses them.

Thanks to all of the publishers who provided review copies of their related materials, and to the authors who answered my electronic mail asking for more detailed descriptions. For anyone else planning to write a large annotated bibliography, I recommend the database program *Bibliographica* by Olaf Winkelhake, which you can download and buy online inexpensively.

My thanks to Marc Bekoff for writing the Foreword and to Emily Birch, Betty Pessagno, and other Greenwood editors for answering all my questions on this major project. Appreciation also to Jack Delivuk for continued encouragement on my writing ventures.

Introduction: How to Use This Book

Animal rights has become one of the most controversial issues of our century and promises to continue as such into the next millennium. Ignored and belittled by most philosophers and teachers throughout history, the subject of animal rights surged into the public consciousness in the nineteenth century. The Industrial and Scientific Revolutions brought many wonderful things to human societies but frightening new things to the animal world: vivisection first, and factory farming later. Gradually, the number of writings about animal rights grew, until the explosion of the 1970s after Peter Singer's groundbreaking work, *Animal Liberation*.

Charles Magel and Stephen Kellert compiled some excellent bibliographies of animal rights literature in the mid-1980s. Most of these were exhaustive, but not annotated: simple book/article listings in alphabetical order. These are useful to be sure, but mainly for scholars. Only doctoral students would wish to plow through 3,800 entries, relying solely on their titles for guessing what these items might be about.

Annotated bibliographies use edited notes to give the readers more help in determining the usefulness, subject, intended audience, and scope of the items in question. *Animal Rights: A Subject Guide, Bibliography, and Internet Companion* contains more than 900 annotated entries, making it the largest bibliography of its kind yet published.

One problem that arises with a long bibliography is the cumbersome nature of searching through the alphabetical lists to find what you want. To reduce this sort of frustration, I have divided the bibliography into six large subject areas: general works, animal natures, fatal uses of animals, nonfatal uses of animals, animal populations, and animal speculations. These subject areas encompass the whole of the animal rights debate, and any given resource can be placed into one of these categories. Because of these subject-area groupings, the researcher can simply determine which topics are of interest, and then look at the appropriate chapter to find related materials. For instance, a student seeking information about the animal rights views on horse racing would look in Chapter 4, "Nonfatal Uses of Animals." Someone seeking books about endangered species would use Chapter 5, "Animal Populations." While looking through a chapter, the researcher will encounter other related materials that may stimulate new ideas. My editorial notes and choice of Internet websites are intended to spur thoughts that will help you to find the information you may need.

I chose items for inclusion in the bibliography based on the following principles:

1. A few foundational works, regardless of age (like Darwin, Salt, and Gompertz).
2. All adult works since 1985, when the last significant bibliographies appeared.
3. A sampling of young adult animal rights materials.
4. A few representative works by prolific authors (Singer, Regan, Linzey).
5. A sampling of available videos.
6. Some older or rare works that alone touch on a relevant subject.
7. Avoidance of most self-published works, which would be difficult to obtain.

While compiling the bibliography, I used the Internet to locate many book reviews and author pages. Originally, the plan was to include an appendix of Internet sites. However, it became apparent that the sites would be much more useful if placed with the relevant entries, so the researcher would not have to flip from chapters to appendices. Readers will find more than 1,200 Internet URLs among the entries. URL addresses are incredibly picky about spelling, so I have put the addresses in brackets. Type in every character you see inside the brackets, but not including the brackets. You should also understand that the Internet is a rapidly changing realm and that addresses are not necessarily permanent. Many of these may be gone by the time you read this book, so do not become frustrated if you encounter missing Internet links. Later in this chapter I will provide some hints on how to find things on the Internet.

Each entry contains the following information, when it was available: author(s), title, edition, series title, location of publisher, publisher, number of pages, year of publication, and International Standard Book Number (ISBN). In the notations, I have included the most important information available to help the research, in the following priority order:

1. An Internet website containing the fulltext of the book, when available.
2. An Internet website containing excerpts or chapters of the book.
3. Internet websites with articles or interviews by the author(s).
4. Short quotations from the book.
5. Internet websites with brief biographies or webpages about the author(s).
6. Internet websites with related articles or links.
7. Short descriptions or summaries of the book that I gleaned personally.

8. Editorial comments, intended to spur thoughts about the issues in the item.

For those of you who are not expertly familiar with Internet searching, allow me to make a few suggestions, as you prepare to use this book. Let's say that you wish to find more Internet references than I have provided or perhaps those addresses have vanished from the Net. First of all, the article may simply have been renamed, this happens a lot. First thing to do is simply hack off the last bit of the address. For instance, if my address listing said [www.nal.usda.gov/awic/pubs/oldbib/qb9329.htm], try backspacing so it says [www.nal.usda.gov/awic/pubs/oldbib/]. You will probably get a "root" directory, which may have a list of available resources. Maybe they ditched the old one in favor of a newer one. If the searches through root directories do not get you the information you want, try my preferred search engine, Alta Vista, found at [www.altavista.com]. Alta Vista has the largest database from which to search. Click on Advanced Search, then use the large white box to the right of "boolean query." Type in the author's full name, the word "and," then the main title (for example, Carol J. Adams and neither man nor beast). This search method is not "case sensitive," so you needn't worry about capital letters. Click Search. If you get 100 or more "hits," you will wish to narrow the search down; add "and" and a subtitle, or some other detail you are interested in, like "cruelty." If you do not get enough hits, start taking away some of the author's name, the initials or first or middle name (never the last name) [e.g., adams and neither man nor beast].

Another way of doing this is by subject searching. Try typing in "feminism and animal rights." If you get too many hits, add words (feminism and animal rights and endangered species). Yahoo is even better than Alta Vista for subject searches. The Yahoo homepage is [www.yahoo.com]. Most of their animal rights subjects can be found at [dir.yahoo.com/Science/Biology/Zoology/Animals_Insects_and_Pets/Animal_Rights/]. From here, you will find many links to organizations and articles on subjects that may help.

Another way to find articles and links that may help you is to simply look at other entries in the same chapter. For instance, if you wanted articles about feminism and animal rights, but the links under Carol J. Adams don't work, simply continue on looking through the chapter. You will later find other feminist writers, like Lynda Birke or Josephine Donovan. Maybe those links will work, or you can use the above search steps using those works.

Also, keep in mind that Chapter 1 of this book, "Animal Rights: General Works," contains resources that try to cover many aspects of the animal rights movement. In other words, those books may have a chapter or two on the subject you want, even if you cannot find a whole book on that subject.

Lastly do not forget your local libraries. Most libraries have trained professionals who can help you find what you are looking for. Many of these librarians have taken classes on Internet searching, or at least Reference searching. More important, many libraries have an Interlibrary Loan service. This means that they may be able to get you some of the books you are interested in, at little or no cost. Simply show them the information you find here, like author, title, and year, and they may be able to borrow a copy from some other library.

I hope that you find this book useful, and I wish you well in your search.

Chapter 1

Animal Rights: General Works

This is the central chapter of the book, which all researchers should browse, even if they intend to focus on one of the more specific subject chapters. It contains the widest variety of animal rights books because it is the "general works" chapter; items that do not fit easily into the other categories are included here. These include bibliographies and encyclopedias, histories, and general philosophical resources. Here are the types of works included in Chapter 1, numbered for easier reference.

1. *Bibliographies, handbooks, and general discussions*. Examples of these works would be Marc Bekoff and Carron Meaney's *Encyclopedia of Animal Rights and Animal Welfare;* Amy Blount Achor's *Animal Rights;* and Peter Singer's *Animal Liberation*. These items cover several issues, not just one, and thus they cannot be placed into the narrower subject categories.
2. *Animal rights law*. These are books pertaining to animal cruelty laws. Works that are very specific, say, for protecting endangered species, would be found in Chapter 5, the "Animal Populations" chapter of this book.
3. *Animal rights histories*. Works that cover the growth of the animal rights movement in various countries and over different time periods are found in this chapter. Histories are necessarily broad in scope, as they touch upon many issues, like vivisection, cruelty, zoos and menageries, philosophies, and so on.
4. *Classic works*. Books that had a major impact on the animal rights movement are included, even when they are ancient. Such works are called "seminal." Henry Salt, Lewis Gompertz, and Albert Schweitzer are among these seminal resources.
5. *Organizations and activists*. Biographies, autobiographies, and animal rights organizational histories are included here. This includes "extremist" works like those of Screaming Wolf, David Foreman, and Paul Watson; plus the "extremist"

critiques of these activists. Some anti-animal rights spokesmen may be as radical as those whom they castigate.

6. *Politics and media*. Books which analyze the impact of animal rights "crusades" on political life and media coverage are included in this chapter. Greenpeace, PeTA, and other such organizations use political influence and advertising very well. Robert Garner has written several of these political gauges of animal rights campaigns.
7. *Ideologies and philosophies*. Books that discuss the many facets of animal rights ideas and philosophies are included here. Such ideas have broad importance to the movement and cannot be categorized into one subject heading. These are usually written by philosophers like Peter Singer or Tom Regan.
8. *Attitudes and motivations*. Resources that ask questions like "Why do people feel this way about animals?" or, "Does animal cruelty lead to human cruelty?" are motivational, and touch directly upon every aspect of the animal rights debate.

---

1. Achor, Amy Blount. *Animal Rights: A Beginner's Guide: A Handbook of Issues, Organizations, Actions and Resources*, 2nd revised ed. Yellow Springs, OH: Writeware, 452 pp., 1996. (0963186515) Covers the whole range of issues in animal rights including: philosophies, organizations, answers to common questions, and sources for obtaining "cruelty-free" products. A good starting place for folks who are new to animal rights is [animalrights.miningco.com/msub101.htm].

2. Adams, Carol J. *Neither Man nor Beast: Feminism and the Defense of Animals*. New York: Continuum, 272 pp., 1995. (0826408036) This author has written several books on "patriarchal" culture, asserting that our civilization has institutionalized the domination of both women and animals. Her essays promote an ideal of female and animal freedom; she also discusses vegetarianism and eco-feminism. Reviewed at [www.psyeta.org/sa/sa4.1/baker.html].

3. Adams, Carol J., and Donovan, Josephine, eds. *Animals and Women: Feminist Theoretical Explorations*. Durham, NC: Duke University, 392 pp., 1995. (0822316676) Essays by Lynda Birke, Gary Francione, and Marian Scholtmeijer, among others. Because women make up a majority of animal rights supporters, feminism has a larger than average influence on the movement. Read the table of contents at [www.geocities.com/Wellesley/8385/wa.html]. See the author's page at [www.triroc.com/carol adams/] and an organization of "Feminists for Animal Rights" at [arrs.envirolink.org/ far/].

4. Allen, John Timothy, compiler. *Ethical and Moral Issues Relating to Animals*. Quick Bibliography Series QB93-29. Beltsville, MD: National

Agricultural Library, 1994. A bibliography that can be accessed at [www.nal.usda.gov/awic/pubs/oldbib/qb9329.htm].

5. Appleby, Michael C., and Hughes, Barry O., eds. *Animal Welfare*. Wallingford: CAB International, 316 pp., 1997. (0851991807) Selected readings on how to keep captive animals healthy. There are discussions on farm animals, veterinary medical care, and psychological aspects of captivity. Solutions are offered to common problems, and there are specific assessment criteria given for determining animal needs. See Appleby's webpage at [helios.bto.ed.ac.uk/ierm/people/academic/appleby.htm].

6. Appleby, Michael C. *What Should We Do About Animal Welfare?* Ames: Iowa State University Press, 200 pp., 1999. (0632050667) "It is a rational examination of the politics and economics of animal welfare, and the single source reference to historical and contemporary scientific and philosophical work in the area to date." The author teaches farm animal behavior at the University of Edinburgh.

7. Arluke, Arnold, and Sanders, Clinton. *Regarding Animals*. Animals, Culture, and Society series. Philadelphia: Temple University, 256 pp., 1996. (1566394406) Two sociology professors interviewed and observed people who worked in animal euthanasia, vivisection, and other animal-use trades, to understand the thinking and rationalizations behind their work. The analysis marvels at the contradictory way in which society treats animals. The Nazi Germans, for instance, were very affectionate with their animals, while killing humans by the truckload. People are able to compartmentalize themselves from their actions upon animals to express human power, control, and superiority. This book is fairly academic but has many deep insights and a good bibliography. Read an article by Arluke at [www.numag.neu.edu/9709/FTheF997.html].

8. Ascione, Frank R., and Arkow, Frank, eds. *Child Abuse, Domestic Violence, and Animal Abuse: Linking the Circles of Compassion for Prevention and Intervention*. West Lafayette, IN.: Purdue University, 380 pp., 1999. (1557531420) Original essays by leaders in the field of psychology. The basic premise is that animal abuse is a frequent indicator of future human and family abuse. Cruel children often become dangerous adults, and they need counseling at a young age ("early intervention"). Read an article by one of the authors at [www.vachss.com/mission/guests/ascione_2.html] and related statistics at [www.americanhumane.org/cpfactviol.html].

9. Baker, R. M., Mellor, D. J., and Nicol, A. M., eds. *Animal Welfare in the Twenty-First Century: Ethical, Educational and Scientific Challenges*.

London: Universities Federation for Animal Welfare, 129 pp., 1994. (095-9054065) Twenty-three papers from a conference at Christchurch, New Zealand. See the list of speakers and topics at [www.adelaide.edu.au/ANZCCART/AnimalWelfare.html].

10. Baker, Steve. *Picturing the Beast: Animals, Identity, and Representation.* New York: St. Martin's, 242 pp., 1993. (0719033780) How animals are depicted in the media, and how these media depictions influence our thinking. One of the more obvious examples of biased media depiction of animals is that of Walt Disney films. Bambi (and myriad spin-offs) show all the animals as equally intelligent and having identical personalities to those of humans. These films have an impact on children; the question is, does this impact wear off with age? Read a related article at [www.sunderland.ac.uk/~as0cba/miners].

11. Barton, Miles. *Why Do People Harm Animals?* Let's Talk About series. New York: Gloucester Press, 32 pp., 1988. Written for juvenile readers. (0531171353) Sadly, this question of motivation in animal cruelty goes largely unnoticed. It has long been known that cruelty to animals is a frequent precursor in a disturbed individual to his or her harming of people. Cruelty to animals is also a largely unenforced crime, and so it can be practiced without any real consequences. Read an article on attitudes toward animals at [www.psyeta.org/sa/sa3.2/driscoll.html].

12. Bekoff, Marc, and Meaney, Carron A., eds. *Encyclopedia of Animal Rights and Animal Welfare*. Westport, CT: Greenwood Publishing, 446 pp., 1998. (0313299773) More than 170 essays in alphabetical order (by subject) which overview all the key issues of animal rights; written by 125 expert contributors, including: Arnold Arluke, Stephen Clark, Michael Fox, Gary Francione, Jane Goodall, Andrew Linzey, Tom Regan, Bernard Rollin, and Peter Singer. This book should be in every library for reference. Read an article by Bekoff at [earthsky.worldofscience.com/Features/News/jinxed-lynx.html]. Reviewed at [www.api4animals.org/Publications/Animal Issues/1999-Spring/Reviews99-01.htm].

13. Birke, Lynda. *Feminism, Animals and Science: The Naming of the Shrew.* Philadelphia: Open University, 167 pp., 1994. (0335191975) "My central theme throughout the book is to explore the meanings of 'animal' and 'human' in relation to the feminist literature on science." (p. 14) "[O]ur cultural beliefs that we are somehow different from — superior to — the mindless morons we construct as 'animals.' " (p. 15) The author is a biologist and feminist from the University of Warwick. Read an article by Birke at [www.psyeta.org/sa/sa3.1/birke.html] and see a biography at [www.cddc.vt.edu/feminism/Birke.html].

14. Blackman, D. E., et al., eds. *Animal Welfare & the Law*. Cambridge: Cambridge University, 283 pp., 1989. (0521364574) Some of the papers presented at the 1986 "Cardiff" seminar. There are laws against cruelty to animals, but they are rarely enforced, and they are generally meaningless misdemeanors when enforced. Why? Is it because the police are overworked, or do not understand the laws? Or because the judges won't enforce the law when brought to trial? Or because the lawyers won't prosecute? What good are laws if unenforced?

15. Brebner, Sue, and Baer, Debbie. *Becoming an Activist: PeTA's Guide to Animal Rights Organizing*. Washington, DC: People for the Ethical Treatment of Animals, 70 pp., 1989. A very practical booklet on creating leaflets, working with the media, raising funds, and other aspects of activism. Brebner is the education director for PeTA, and a registered nurse. See a site on how to start your own group at [www.enviroweb.org/animal_watch/start.html]. Read a brief interview with Brebner at [www.how-on-earth.org/backissues/10_Sue.html].

16. Briggs, Anna C. *For the Love of Animals: The Story of the National Humane Education Society*. Leesburg, VA: Potomac, 122 pp., 1990. Short descriptions on the different kinds of problems that are encountered by the National Humane Education Society. A few of the topics included are: fur, overpopulation, vivisection, and animals used in entertainment. Read an article by the author at [www.atbeach.com/veggie/articles/ pledging.html].

17. Brooman, Simon, and Legge, Debbie. *Law Relating to Animals*. London: Cavendish, 462 pp., 1997. (1859412386) Two British law professors suggest some possible changes for legal action in the twenty-first century. Read about dog laws at [www.canismajor.com/dog/laws1.html] and [www.canismajor.com/dog/laws2.html]. Traditionally, the laws regarding animals had to do only with animals as property. Anti-cruelty laws had nothing to do with saving animals from pain but with saving the owners from emotional pain due to the actions of another person.

18. Caras, Roger A. *A Perfect Harmony: The Intertwining Lives of Animals and Humans Throughout History*. New York: Simon & Schuster, 272 pp., 1996. (0684835312) Caras, the president of the American Society for the Prevention of Cruelty to Animals, is called "America's most celebrated writer on pets and wildlife." This book is a history of domestication. Read the table of contents at [www.earthfriendlybooks.com/Sections/catalog/nature/0684835312.html].

19. Carruthers, Peter. *The Animals Issue: Moral Theory in Practice*. Cambridge: Cambridge University, 206 pp., 1994. (0521436893) This philo-

sopher uses "contractualism" to argue that animals do not have rights. He writes that animals do not possess consciousness; however, humans should extend some benefits of social contract to animals just out of human compassion. Read a review of a similar work by Carruthers at [psyche.cs.monash.edu/v5/psyche-5-16-saidel.html]. See the author's page at [www.shef.ac.uk/uni/academic/N-Q/phil/department/staff/Carruthers.html] and read an article by the author at [boromir.cs.monash.edu.au/v4/psyche-4-03-carruthers.html].

20. Carson, Gerald. *Men, Beasts, and Gods: A History of Cruelty and Kindness to Animals.* New York: Scribners, 268 pp., 1972. (684130394) A history of societies' attitudes toward animals: including animism and worship, arenas and sports, experimentation, etc. "This book is about the relationships that have existed between mankind and the lower animals from those prehistoric eras we have to guess about down to the present time. It is also about the ideas that men have articulated to explain their attitudes." (p. vii) Related links at [arrs.envirolink.org/ar-issues/].

21. Chadwick, Ruth, ed. *Encyclopedia of Applied Ethics.* New York: Academic, 3,101 pp., 1997. (0122270657) An outstanding compilation of essays on all ethical problems. Look up "Animals" to find cross-references to articles on major animal rights issues. Many large college and university libraries own this reference work.

22. Clark, Stephen R. L. *Animals and Their Moral Standing.* New York: Routledge, 208 pp., 1997. (0415135605) Major essays on the historical development of animal rights in theory and practice. He says that it is wrong to cause avoidable suffering. Read an interview with the author at [arrs.envirolink.org/ar-voices/three_aw.html]. See the author's resume at [www.liv.ac.uk/~srclark/stephen.html].

23. Clarke, Paul A. B., and Linzey, Andrew, eds. *Political Theory and Animal Rights.* London: Pluto, 193 pp., 1991. (0745303862) Essays by leaders in the animal rights movement.

24. Clough, Caroline, and Kew, Barry. *The Animal Welfare Handbook.* London: 4th Estate, 1993. Major arguments for animal rights are discussed, and a directory of groups is included. See some good links at [www.urmc.rochester.edu/ucar/links.html].

25. Coatzee, J. M. *The Lives of Animals.* University Center for Human Values series. Princeton, NJ: Princeton University, 130 pp., 1999. (0691004439) A semi-fictional work about animal cruelty and vegetarianism that has

generated recent attention. The author is an English professor from South Africa.

26. Cooper, Margaret E. *An Introduction to Animal Law.* New York: Academic, 213 pp., 1988. (0121880303) Reviewed at [www.nsplus.com/nsplus/insight/animal/law.html]. Find links to animal law sites at [world animal.net/wan/useful.html].

27. Crisp, Terri, and Glen, Samantha. *Out of Harm's Way: The Extraordinary True Story of One Woman's Lifelong Devotion to Animal Rescue.* New York: Pocket, 393 pp., 1996. (0671522787) The exciting stories of how one woman has been saving animals from natural and man-made disasters (like the Exxon Valdez oil spill). Two sites on keeping your pets safe during disasters are [www.uan.org/programs/ears/companion.htm] and [www.icpost.com/columns/health/meier/dvm0398.htm]. There are critics of this sort of action. While perhaps admitting that the sentiment of saving animals is noble, many have the idea that we are "wasting" time, money, and effort trying to save these animals when there are far more important missions, like saving humans, to be accomplished. Is it stupid to enter a burning house to save a pet? It is "just an animal," or is it? Is it a member of that family, with more value than that of a simple pet?

28. Dale, Stephen. *McLuhan's Children: The Greenpeace Message and the Media.* Toronto: Between the Lines, 192 pp., 1996. (1896357040) One of the founders of Greenpeace, Paul Watson, says that Greenpeace was "the first organization to understand the nature of the media." This book describes how deliberate and coordinated use of the media has led to major successes for this animal rights/environmental organization. It includes interviews with activists and media specialists, and speculation on how the "fragmentation of the mass media" may affect this media success. Do organizations like Greenpeace put image over action? Have a look at the extent of Greenpeace's use of publicity at [www.greenpeace.org/press releases/]. Another question beyond that of the extent of media use is that of the accuracy and purpose of the messages being distributed. Critics claim that such organizations receive millions of dollars in donations but use it all on congressional lobbying and media commercials, not on real efforts to save the animals. And how often is the message slightly skewed, implying untruths to sensationalize the story and improve donations? Reviewed at [www.esemag.com/0197/0197ed.html].

29. Datta, Ann, ed. *Animals and the Law: A Review of Animals and the State.* Otter Memorial Paper, No. 10. Chichester, England: Chichester Institute of Higher Education, 104 pp., 1998. (0948765690) Includes articles by Andrew Linzey and Richard D. Ryder.

30. DeRose, Chris. *In Your Face: From Actor to Animal Activist.* Los Angeles: Duncan, 303 pp., 1997. (0965321908) There is a web page by the author about this book at [www.lcanimal.org/book.htm]. The former actor describes how he fights pet thievery, which he says the USDA (U.S. Department of Agriculture) promotes by issuing of class B licenses to dealers. Reviewed at [www.vegsource.org/books/face.htm]. See a list of some other famous pro-animal rights people at [www.traverse.com/people/bowhunter/antis.html].

31. Dolins, Francine L., ed. *Attitudes to Animals: Views in Animal Welfare.* Cambridge: Cambridge University, 262 pp., 1999. (052147342x) Seventeen essays about human views on animals, including works by James Serpell and Mary Midgley. What is it to be a human or an animal? The author works with the Humane Society of the United States (HSUS). See a biography of the author at [www.hsus.org/programs/research/francine.html].

32. Dombrowski, Daniel A. *Babies and Beasts: The Argument from Marginal Cases.* Chicago: University of Illinois, 216 pp., 1997. (0252023420) The uses and abuses of "marginal cases" in philosophy, which include the treatment of infants, comatose patients, and retarded humans. The author shows that such marginal cases are no different than animals, and thus animals should be treated by a similar standard. He examines the opinions of Peter Singer, Tom Regan, Raymond Frey, and Peter Carruthers. See a brief description at [www.ctr4process.org/ABOUT%20CPS/SEMINARS/Sem67.htm].

33. Dombrowski, Daniel A., and Tiger, Steve, eds. *Hartshorne and the Metaphysics of Animal Rights.* Albany: State University of New York, 159 pp., 1988. (0887067042) Dombrowski is a philosophy professor at Seattle University. See a biography of Hartshorne at [www.ctr4process.org/WHATISPRCS/Hartshorne.htm].

34. Donovan, Josephine, and Adams, Carol J., eds. *Beyond Animal Rights: A Feminist Caring Ethic for the Treatment of Animals.* New York: Continuum, 216 pp., 1996. (0826408362) Eight essays that promote the ethic of "care" as a basis for the treatment of animals for feminists. See Donovan's bibliography at [www.umaine.edu/english/facultypages/donovanpubs.htm]. Read an article about the feminist presence in the animal rights movement at [animalnews.com/news/ malepractice.htm].

35. Ferguson, Moira. *Animal Advocacy and Englishwomen, 1780-1900: Patriots, Nation, and Empire.* Ann Arbor: University of Michigan, 192 pp., 1998. (0472108743) A critical look at British women writers who

expressed their moral outrage at animal abuses, which allowed them to "achieve a public voice as citizens." Both conservative and liberal, pro and con, are represented among the writers. See the brief publisher review at [www.press.umich.edu/titles/10874.html].

36. Field, Shelly. *Careers as an Animal Rights Activist.* New York: Rosen, 205 pp., 1993. (0823914658) This is a young adult book, however, it is the only work of its kind of which I am aware. It includes both paying and volunteer opportunities in the field. Find information about careers in lab animal veterinary science at [www.aslap.org/career.htm]. See a neat site with lots of animal careers at [www.arkanimals.com/Career/Career1.html].

37. Finsen, Lawrence, and Finsen, Susan. *Animal Rights Movement in America: From Compassion to Respect.* Boston: Twayne, 309 pp., 1994. (0805738843) Profiles the history of animal rights and many of the groups involved in the debate. The Animal Protection Institute says this is "probably the best book written on all aspects of the moral crusade of animal rights." Read an article by Susan Finsen at [www.lclark.edu/~alj/intro3.html].

38. Foreman, David. *Confessions of an Eco-Warrior.* New York: Crown, 228 pp., 1993 (originally published in 1991). (051788058x) An "ecological call to arms" by a former leader of Earth First! The author is known for promoting extreme tactics, such as the intentional destruction of logging machines, in seeking to protect the environment. In this book he says, "the only hope of the Earth is to withdraw huge areas as inviolate nature sanctuaries from the depredations of modern industry and technology," and "We humans have become a disease, the Humanpox." See a guide to direct action at [www.enviroweb.org/animal_watch/Rguide1.html].

39. Foreman, David, ed. *Ecodefense: A Field Guide to Monkey Wrenching* (1985), 3rd ed. Chico, CA.: Abbzug Press, 311 pp., 1993. (0963775103) A manual on how to sabotage property (in defense of the environment), which also bears the necessary caveat "for entertainment purposes only." See a biography of Dave Foreman at [www.project21.org/dos7129.htm]. Is there such a thing as "nonviolent" sabotage of property? Earth First and such groups practice this "nonviolent" action. No humans are supposed to be directly harmed. However, occasionally there are accidents. For example, an activist puts metal spikes into a tree to ruin the logger's chainsaw. Sometimes loggers have been injured by the chainsaw's circular blade coming apart when striking the unseen metal spike. Does that make the action violent, or just an accident?

40. Fox, Michael W. *Returning to Eden: Animal Rights and Human Responsibility*. New York: Viking, 281 pp., 1980. (0898749344) The author says that our present distance from the Garden of Eden is measured by the amount of animal abuse in our world. He also proposes that humans "may be a mistake, a tragic error" of evolution, as seen by the destruction mankind has wrought.

41. Fox, Michael W., and Amory, Cleveland. *Inhumane Society: The American Way of Exploiting Animals*. New York: St. Martin's, 268 pp., 1992 (originally published in 1990). (0312042744) How veterinarians are often breaking their vows by abusing animals (because they profit from animal exploitation in industry) in our modern society. Fox writes that U.S. laboratories use between 25 million and 35 million animals each year for research. The book discusses vivisection, wildlife, zoos, pets, genetic engineering and other subjects. He says that animal rights ideas are neither irrational nor solely sentimental. See a brief biography at [petremedies. com/bio.html]. This is an interesting indictment of veterinarians not altogether different from the questions always faced by medical personnel. Is it wrong to help cure a patient when the patient will certainly be abused again? Should the vet help the factory farmer keep down diseases in the barn, though the cattle will soon be slaughtered anyway? Who is the vet more obligated to protect, the farmer or the animal?

42. Francione, Gary L. *Animals, Property, and the Law*. Ethics and Action series. Philadelphia: Temple University, 368 pp., 1995. (1566392845) Famed lawyer William M. Kunstler wrote the foreword. Why does the law not protect animals from exploitation? Read an article by the author at [www.geocities.com/RainForest/2062/five.html]. Our laws protecting animals are not really protecting the creatures because of their intrinsic value but because of their value to humans as property. The crime of cruelty is often measured by the "street value" of the creature, or perhaps its emotional value to the owner, but not because it is believed to have any value in and of itself. This is a foundational problem in the legal system, from an animal rights perspective. Read an article by the author at [www. geocities.com/RainForest/2062/five.html].

43. Francione, Gary L. *Introduction to Animal Rights: Your Child or the Dog?* Philadelphia: Temple University, 264 pp., 1999. (1566396921) An analysis of the strange and dramatic inconsistency of the American public, which claims to oppose animal uses but eats hamburgers, goes to circuses, and buys products that are animal tested. The author attacks the animal welfare movement as harmful to true animal rights. Read an article by the author at [arrs.envirolink.org/ar-voices/rlr_ar_and_aw.html]. See also [www. animal-law.org/library/library.htm].

44. Francione, Gary L. *Rain Without Thunder: The Ideology of the Animal Rights Movement.* Philadelphia: Temple University, 266 pp., 1996. (156-6394619) Why the philosophy of animal welfare cannot lead to animal liberation, and in fact may hurt the animal rights movement. The author says that simple improvements in animal welfare further entrench animal oppression. See two negative reviews at [www.firstthings.com/ftissues/ft9710/derr.html] and [www.naiaonline.org/book9708.html]. It is also reviewed at [www.montelis.com/satya/backissues/may97/tails.html]. This is a common opinion among animal rights people, who are intolerant of animal welfare folks. They say that animal welfare just makes the slavery a little softer, and thus harder to end. This lack of moderation leads to the strengthening of conflicts, however, and makes intermediate solutions almost impossible to negotiate. It's all or nothing.

45. Fraser, Laura, and Horwitz, Joshua. *The Animal Rights Handbook: Everyday Ways to Save Animal Lives.* Los Angeles: Living Planet Press, 113 pp., 1990. (0962607207) How consumers' choices of foods and clothes impact animal lives. Also a list of cruelty-free products. This book was sponsored by the ASPCA.

46. Free, Ann Cottrell. *Animals, Nature and Albert Schweitzer*, revised ed. Washington, DC: Animal Welfare Institute, 93 pp., 1988 (originally published in 1982). Albert Schweitzer (famed explorer and doctor) saw animals in the light of his philosophy "reverence for life." This book can be found free on the internet at [www.animalwelfare.com/schweitzer/as-idx.htm].

47. Freeman, Michael R. *Animal Welfare: Index of Modern Information.* Washington, DC: ABBE, 150 pp., 1990. (1559141875) See a good list of animal welfare organizations at [pwww.vetmed.ucdavis.edu/Animal_Alternatives/organiza.htm].

48. Frey, Raymond Gillespie. *Interests and Rights: The Case Against Animals.* New York: Oxford University, 176 pp., 1980. (0198244215) Charles Magel summarizes the premise of this influential book as "animals have no desires because they have no beliefs because they have no propositions because they have no linguistic capacity." The author says that language is needed for sentience and beliefs and desires. He approaches the arguments from a utilitarian philosophy, just as Peter Singer does, with opposite conclusions. This sort of argument is countered by more recent discoveries in animal ethology. Do animals in fact have no languages? It is more debatable now than it was in 1980.

49. Galer, Eilleen Gardner, and Kroitzch, Gregory J., eds. *God Barking in Church and Further Glimpses of Animal Welfare*. Plymouth, England: Five Corners, 80 pp., 1994. (0962726273) The author worked for the local Humane Society for more than twenty years. She shares some stories that are relevant to animal welfare issues. See a brief biography at [www.fivecorners.com/fcp/eilleen.htm].

50. Garner, Robert, ed. *Animal Rights: The Changing Debate*. Politics, Medical Science, Philosophy series. New York: New York University, 256 pp., 1997. (0814730981) An anthology of essays (including Peter Singer, Michael Fox, Richard Ryder, Gary Francione, and Andrew Rowan) over the last 25 years, showing the reasons for the rise of radicalism in recent times. The author is an expert in politics.

51. Garner, Robert. *Animals, Politics, and Morality*. Issues in Environmental Politics series. Manchester University, 258 pp., 1993. (0719035759) The author says that zoos are not intrinsically evil, though they are evil if animals are kept in suffering conditions and are there solely for amusement purposes. Reviewed at [www.newscientist.com/nsplus/insight/animal/moral.html].

52. Garner, Robert. *Political Animals: Animal Protection Policies in Britain and the United States*. New York: St. Martin's, 286 pp., 1998. (0312212089) Comparative research on animal welfare policies and how special interest groups bring influence to bear on the governments involved. Read an article by the author at [www.psyeta.org/sa/sa3.1/garner.html].

53. Gleason, Sean J., and Swanson, Janice C., compilers. *An Annotated Bibliography of Selected Materials Concerning the Philosophy of Animal Rights*. Beltsville, MD: National Agricultural Library, 17 pp., 1988. See a nice bibliography on the subject at [www.enviroweb.org/animal_watch/Bookab.html].

54. Godlovitch, Stanley, Godlovitch, Rosalind, and Harris, John, eds. *Animals, Men and Morals: An Inquiry into the Maltreatment of Non-Humans*. New York: Taplinger, 238 pp., 1971. (0394178254) Thirteen essays on fur, sports, meat, factory farming, entertainment, and so on. "Once the full force of moral assessment has been made explicit there can be no rational excuse left for killing animals, be they killed for food, science, or sheer personal indulgence." (p. 7) Peter Singer says that this book is the manifesto for animal liberation.

55. Gold, Mark. *Animal Century: A Celebration of Changing Attitudes to Animals*. Northvale: Jon Carpenter, 240 pp., 1999. (1897766432) I have not been able to obtain a copy of this for review.

56. Gold, Mark. *Animal Rights: Extending the Circle of Compassion*. New York: Oxford University, 153 pp., 1995. (1897766165) How ordinary people can run campaigns and use "caring consumerism" to bring success to the animal rights cause. See the author's website at [www.holisticmed.com/].

57. Gompertz, Lewis. *Moral Inquiries on the Situation of Man and Brutes*. Mellen Animal Rights Library #H2. Lewiston, NY: Edwin Mellen, 204 pp., 1997 (originally published in 1824). (0773487220) A reprint of an early nineteenth-century vegetarian book. The author, a pioneer of the Royal Society for the Prevention of Cruelty to Animals (RSPCA), says that the purpose of morality is to make all beings happy, including animals. The editor, Charles Magel, says that Gompertz' ideas are "stronger" than those of Henry Salt or Peter Singer. Read three classic articles by Gompertz at [arrs.envirolink.org/ar-voices] in the farmed animals section.

58. Goodall, Jane and Berman, Phillip. *Reason for Hope: A Spiritual Journey*. New York: Warner, 320 pp., 1999. (0446522252) This is an autobiography of the famous scientist who has been studying chimpanzees in Africa for more than thirty years. For the last decade, she has been traveling the world to spread the message of ecology to the public. As the title implies, the book includes her personal and spiritual struggles (and victories). This was made into a one hour PBS special, which can be purchased on video. It is refreshing to see that not all scientists have abandoned religion or personal ethical value systems.

59. Guither, Harold D. *Animal Rights: History and Scope of a Radical Social Movement*. Carbondale: Southern Illinois University, 272 pp., 1998. (080-9321998) The writer is a professor of Agricultural Policy at the University of Illinois, and a moderate who offers an objective look at the animal rights movement. He gives an "unbiased examination of the paths and goals of the members of the animal rights movement and of its detractors." (from the dustjacket) Read an article by the author at [ianrwww.unl.edu/farmbill/aniright.htm]. See links of anti-animal rights sites and articles at [www.dandy-lions.com/animal_rights.html].

60. Ham, Jennifer, and Senior, Matthew, eds. *Animal Acts: Configuring the Human in Western History*. New York: Routledge, 258 pp., 1997. (04159-16097) A scholarly analysis of the views of several philosophers, including

Rene Descartes. Read a brief review of the book at [www.montelis.com/satya/backissues/sep97/remembering_bobby.html].

61. Hardy, David T. *America's New Extremists: What You Need to Know About the Animal Rights Movement.* Washington, DC: Washington Legal Foundation, 53 pp., 1990. An Arizona lawyer who specializes in wildlife and environmental law makes extensive comments on the animal rights movement. See a brief biography at [www.indirect.com/www/dhardy/antigovernment.html]. Read an anti-animal rights article at [www.aynrand.org/objectivism/animals.html].

62. Harnack, Andrew, ed. *Animal Rights: Opposing Viewpoints*, reprint ed. San Diego: Greenhaven, 240 pp., 1996. (1565103998) Written for young adults.

63. Healey, Kaye. *Animal Rights.* Wentworth Falls, Australia: Spinney, 45 pp., 1998. (1876811188) An Australian view on the differences between animal welfare and animal rights, farming techniques, vivisection, and other issues. See the RSPCA Australia site at [www.rspca.org.au].

64. Hendrick, G., and Hendrick, W., eds. *The Savour of Salt: A Henry Salt Anthology.* Fontwell, England: Centaur, 204 pp., 1989. (0900001305) Salt was probably the greatest of the nineteenth-century animal rights thinkers (1851-1939). He was the first to title a book Animal Rights. Read a biography and quotations at [arrs.envirolink.org/ar-voices/H_Salt2.html].

65. Henshaw, David. *Animal Warfare: The Story of the Animal Liberation Front.* London: Fontana, 1989 (originally published in 1984). The rise of direct action tactics for promoting the philosophy of animal rights. The Animal Liberation Front has become known in recent years for breaking into laboratories, damaging the facilities, and releasing the animals. The FBI considers them to be terrorists. See an anti-direct-action site at [www.terrorism.net/Pubs/csisarticles/com21e.htm]. See a pro-Animal Liberation Front site at [hedweb.com/alffaq.htm].

66. Hersovici, Alan. *Second Nature: The Animal Rights Controversy.* Montreal: CBC, 255 pp., 1985. (0887941494) Animal rights critiqued from an environmentalist and cultural perspective. Animal rights groups are forcing their ideas on natives in many countries; is this cultural imperialism? Originally this was a three-part radio show. Read an article by the author at [www.furcommission.com/resource/perspect93.html].

67. Hicks, Esther K., ed. *Science and the Human-Animal Relationship*. Amsterdam: SISWO, 242 pp., 1993. I have not been able to find any other information about this work.

68. Hoage, Robert J., ed. *Perceptions of Animals in American Culture*. National Zoological Park Symposia for the Public. Washington, DC: Smithsonian, 151 pp., 1989. (0874744938) Papers on animals in civilization; social aspects of how peoples' perceive other creatures. Read an article by the author at [www.si.edu/natzoo/zooview/animals/perceptn.htm].

69. Hogan, Linda, et al., eds. *Intimate Nature: The Bond Between Women and Animals*. New York: Fawcett Columbine, 480 pp., 1998. (0449003000) Seventy-seven stories by women on their love and appreciation for animals. Reviewed at [www.api4animals.org/Publications/AnimalIssues/1998-Summer/Reviews98-02.htm] and [www.webmist.com/review/q2_1998.htm#Intimate]. See a brief biography and some links at [www.hanksville.org/storytellers/linda/].

70. Howard, Walter E. *Animal Rights vs. Nature*. Davis, CA: W. E. Howard, 229 pp., 1990. (0962764108) The author has a Ph.D. in vertebrate ecology and is a Professor Emeritus at the University of California at Davis. Some reviewers blast this book as horribly misrepresenting the animal rights philosophies. The author writes, "What is morally wrong with exploiting animals for human purposes, if done humanely? All animals exploit other animals." (p. 1) This book "discusses the serious dangers of the current extreme animal rights movement." (back cover)

71. Huebotter, Carol. *Teaching Humane Education: A Resource Guide for Staff and Volunteers* (workbook and video). Hinsdale, CO: Hinsdale Humane Society, 35 pp., 1996. Includes a sixty-minute training video. It was written by a schoolteacher who is also a humane education director. See the homepage at [www.enteract.com/~hhs/humedpkg.html].

72. Humane Society of the United States (HSUS). *Making a Difference for Animals: A Look at the Humane Society of the United States*. Washington, DC: Humane Society, 123 pp., 1996. Stories and reports on the importance and goals of the HSUS, whose main website is at [www.hsus.org]. Read an anti-HSUS article at [www.capitalresearch.org/ ap/ap-1097.html]. Some critics say that the HSUS has joined the more radical wing of activists.

73. Hunt, Mary, and Juergensmeyer, Mark. *Animal Ethics: An Annotated Bibliography*. Berkeley: Graduate Theological Union, 1977.

74. Hunter, Robert. *Warriors of the Rainbow: A Chronicle of the Greenpeace Movement.* New York: Holt, Rinehart & Winston, 454 pp., 1979. (0030437415) The author founded Greenpeace in 1970 and was its president. The Greenpeace homepage is at [www.greenpeace.org]. See an anti-Greenpeace site at [luna.pos.to/whale/gen_art_green.html]. See a brief author biography at [www.estreet.com/orgs/sscs/people/hunter.html].

75. Jasper, James M., and Nelkin, Dorothy. *The Animal Rights Crusade: The Growth of a Moral Protest.* New York: Free Press, 214 pp., 1992. (0029161959) Sociologists give a balanced analysis of the animal rights movement, including the issues of vivisection, fur, zoos, wildlife, and lobbying. Reviewed at [www.psyeta.org/sa/sa1.1/takooshian.html].

76. Jasper, Margaret C. *Animal Rights Law.* Legal Almanac series. New York: Oceana, 158 pp., 1997. (0379112442) See a related site at [ipl.unm.edu/cwl]. See the page of the Animal Legal Defense Fund at [www.aldf.org].

77. Jenkins, Sid, and Leitch, Michael. *Animal Rights and Human Wrongs.* Harpenden, England: Lennard, 207 pp., 1992. (1852911050) An RSPCA officer recounts his experiences, which often demonstrate the lack of adequate animal legislation in England. "There is no doubt in my mind that zoos are a thing of the past and should be phased out. Animals do not belong behind bars." See the RSPCA site at [www.rspca.org].

78. Kalechofsky, Roberta. *Autobiography of a Revolutionary: Essays on Animal and Human Rights.* Marblehead, MA: Micah, 189 pp., 1991. (091628834x) Fourteen essays. This author is Jewish and has several works on mixing Judaism and vegetarianism (see her entries in the Animal Speculations chapter of this book). "[A]nimal abuse remains institutionalized with powerful economic and political entanglements. Thus, it resembles slavery rather than child abuse." (p. v) Biography at [www.micahbooks.com/speaker. html].

79. Kean, Hilda. *Animal Rights: Political and Social Change in Britain since 1800.* London: Reaktion, 248 pp., 1998. (1861890141) How animal rights aligned itself with feminists, socialists, and other "radicals" for political and social reform. Histories of campaigns in England, especially regarding vivisection, zoos, hunting, and vegetarianism. This is an outstanding work, with a decade-by-decade analysis of the movement. See a review at [ivu.org/oxveg/Publications/Oven/Books/animal_rights.html].

80. Kellert, Stephen R. *A Bibliography of Human-Animal Relations.* New Haven, CT: Yale University, 1985. (0819149586) 3,861 unannotated citations with no pagination.

81. Larson, Jean A., and Jensen, D'Anna J. B. *Audio-Visuals Relating to Animal Care, Use and Welfare*. AWIC Series 96-02. Washington, DC: Animal Welfare Information Center, 76 pp., 1996. Read the whole work at [www.nal.usda.gov/awic/pubs/awic9601.htm]. See also related works at [arrs.envirolink.org/Faqs+Ref/vadivu/].

82. Lawrence, Elizabeth Atwood. *Hunting the Wren: Transformation of Bird to Symbol: A Study in Human-Animal Relationships*. Knoxville: University of Tennessee, 234 pp., 1997. (0870499602) The complexities of human and animal relationships. The title alludes to the tradition of the wren hunt, which is a bird usually revered but at the time of summer solstice is hunted down. "It is urgent, if we are ever to establish harmonious relationships with other forms of life on the planet, to understand how the beliefs of a particular culture, in combination with observations about the natural history of an animal, result in a symbolic entity that elicits emotional and intellectual responses that influence society's treatment of that animal and may ultimately determine the fate of the species." (p. xii) How does a being become a symbol, and what does that mean to its life and survival? Read an article by the author at [www.psyeta.org/ sa/sa2.2/lawrence.html].

83. Leahy, Michael P. T. *Against Liberation: Putting Animals in Perspective*. New York: Routledge, 273 pp., 1994. (0415103169) The author says that animal rights arguments are unconvincing, and that in human captivity the creatures usually have better lives than they would have experienced in the wild. Animals are primitive beings; fur, vivisection, and hunting are morally right. Reviewed at [freespace.virgin.net/old.whig/fl20anim.htm].

84. Leavitt, Emily Stewart, et al. *Animals and Their Legal Rights: A Survey of American Law from 1641 to 1990*, 4th ed. Washington, DC: Animal Welfare Institute, 441 pp., 1990. (0686278127) The introduction to this book can be seen at [www.animalwelfare.com/pubs/bk-conts.htm#dealers]. Includes legal cases about slaughter, laboratories, educational uses of animals and so on.

85. Lockwood, Randall. *Animal Rights, Animal Welfare, and Human-Animal Relationships: An Annotated Bibliography for Higher Education*. Washington, DC: Humane Society of the United States, 37 pp., 1986. A brief bibliography with annotations of popular works. See a brief biography of the author at [www.puppyworks.com/speakerprofiles.html#anchor135 7830].

86. Lockwood, Randall, and Ascione, Frank R. *Cruelty to Animals and Interpersonal Violence: Readings in Research and Application*. West Lafayette, IN: Purdue University, 452 pp., 1998. (1557531056) These are academic

and professional articles with historical background, case studies, criminology, and psychology, showing the relationship of animal abuse to human abuse. The author is vice president for training at the Humane Society of the United States. Reviewed at [www.api4animals.org/Publi cations/AnimalIssues/1999-Summer/Reviews99-02.htm]. See a related article at [www.abanet.org/child/8-4tip.html].

87. Lutherer, Lorenzo Otto, and Simon, Margaret Sheffield. *Targeted: The Anatomy of an Animal Rights Attack.* Norman: University of Oklahoma, 190 pp., 1992. (080612492x) Very strong attack on the animal rights movement, comparing them to Nazis and terrorists. Instructions for laboratories on how to prepare against attacks of sabotage on facilities. Read a related article at [www.the-scientist.lib.uperm.edu/yr1992/oct/ kaufman_p1_921026.html]. Due to the recent growth of expensive attacks on laboratories, the federal government has passed new, stronger laws in an attempt to deter groups like the Animal Liberation Front. The Federal Bureau of Investigation is now heavily involved in the search for such groups. See Lutherer's faculty page at [www.physiology.ttuhsc.edu/ Lutherer/Lol.htm].

88. Lynge, Finn. *Arctic Wars, Animal Rights, Endangered Peoples.* Hanover: University Press of New England, 118 pp., 1992. (0874515882) The author is a member of the European Parliament as a representative for Greenland; he opposes animal rights in favor of the rights of indigenous peoples to hunt. Read articles by the author at [www.highnorth.no/so-pr-re.htm] and [www.fur.ca/online_resources/ lynge.html]. This is no easy subject. Greenpeace and other antisealing and antiwhaling organizations did not intend to shut down all sealing and whaling; they initially stood only against the nonnative "harvesting" of such animals. But the media and viewers could not understand the distinction; animal death was animal death, regardless of who did the killing. Subsequent legal actions (banning seal furs in many countries) ended up hurting the native peoples a great deal. These natives use animals not just for food but for fuel, clothing, and to trade for other products. It is not a sport to them.

89. Machan, Tibor R. *Do Animals Have Rights?* Social Affairs University, 1990. The author of this essay, a political libertarian, says that animals cannot be moral agents (as people are) and therefore have no rights. The author was a philosophy professor at Auburn University until recently. Read this essay by the author at [genius.ucsd.edu/~john/p/libuniv_dir/ Machan_dir/Machan.AnimRights].

90. Magel, Charles R. *A Bibliography on Animal Rights and Related Matters.* Washington, DC: University Press of America, 602 pp., 1981. (08191-1488x) A comprehensive bibliography with 3,210 entries.

91. Magel, Charles R. *Keyguide to Information Sources on Animal Rights.* Jefferson, NC: McFarland, 267 pp., 1989. (0720119847) More condensed than his earlier bibliography, this work focuses on the origins and philosophy of the animal rights movement. It contains 335 annotated entries.

92. Maguire, Stephen, and Wren, Bonnie, eds. *Torn By the Issues: An Unbiased Review of the Watershed Issues in American Life.* Santa Barbara, CA: Fithian Press, 382 pp., 1994. (1564740935) Chapter four of this book is on animal rights, and is an outstanding summary of basic statistics, chronologies, and issues.

93. Manes, Christopher. *Green Rage: Radical Environmentalism and the Unmaking of Civilization.* New York: Little, Brown, 291 pp., 1990. (0316545139) He defends radical environmentalism and wishes to replace "dominion" with "deep ecology." The author was a Fulbright scholar. Reviewed at [www.visi.com/~contra_m/cm/reviews/cm15_rev_waste.html]. I recently read an interview with this author; he says that he is much more moderate than he used to be.

94. Manning, Aubrey, and Serpell, James, eds. *Animals and Human Society: Changing Perspectives.* London: Routledge, 199 pp., 1994. (0415091551) Ten essays on historical and modern attitudes toward animals. Authors include Arnold Arluke, Murry Cohen, Juliet Clutton-Brock, and Mary Midgley. Reviewed at [www.psyeta.org/sa/sa3.2/sanders.html].

95. Manzo, Bettina. *The Animal Rights Movement in the United States, 1975-1990: An Annotated Bibliography.* Metuchen, NJ: Scarecrow, 306 pp., 1994. (0810827328) The publisher catalog says "Over 1,300 annotated citations that address the animal rights movement's goals, organizations, philosophical underpinnings, and political, educational, and legislative activities between 1975 and 1990."

96. Marquardt, Kathleen. *Animal Scam: The Beastly Abuse of Human Rights.* Washington, DC: Regnery Gateway, 221 pp., 1993. (0895264986) Founder of the group "Putting People First." When her children were taught at school by PeTA (People for the Ethical Treatment of Animals) that she was a murderer because she hunted, she wrote this book. The author cites the most extreme statements and actions of animal rights groups. Her reaction is understandable, since a parent never likes to be called a murderer to her

children. Reviewed at [www.api4animals.org/Publications/AnimalIssues/1994-Spring/Reviews94-01.htm]. See a biography at [www.americanpolicy.org/plate.main/WhoWeAre/KathleenMarquardt.html].

97. Mason, Jim. *An Unnatural Order: Why We Are Destroying the Planet and Each Other*. New York: Continuum, 352 pp., 1997. (0826410286) Describes which Western beliefs lead to racism, sexism, animal cruelty, and domination. The author says that our culture now sees all animals as either being resources for use or pests for killing. Read three brief excerpts at [www.vegan.com/current/jm020298.html]. "Western civilization" gets the blame for a lot of things. In reality, the Western cultures are probably no worse than the Eastern ones. Perhaps the Eastern religions do have more spoken and written lip-service to kindness and peace, but it rarely has worked out that way in practical matters. Animals do not have a good lot in the Eastern nations either.

98. McCulloch, Jennifer. *Creatures of Culture: The Animal Protection and Preservation Movements in Sydney, 1880-1930*. Melbourne, Australia: Melbourne University Press, 256 pp., 1996. (0522846874) The author is a history teacher at Griffith University.

99. McDowell, R. E. *A Partnership for Humans and Animals*. Raleigh, NC: Kinnic Publishers, 95 pp., 1991. (188076203x) I have not found further information on this work.

100. Meyer, Christiane. *Animal Welfare Legislation in Canada & Germany: A Comparison*. European University Studies series 2: Law, vol. 2007. Frankfurt, Germany: Peter Lang, 285 pp., 1996. (3631307330)

101. Midgley, Mary. *Animals and Why They Matter: A Journey Around the Species Barrier*. Athens: University of Georgia, 158 pp., 1983. (08203-20412) The author examines philosophies that try make animals seem unimportant, such as rationalism and anthropomorphism, and is especially critical of the traditional view that reason is the foundation of moral status. See an article by the author at [arrs.envirolink.org/ar-voices/species_on_ice.html] and a brief biography at [www.ivu.org/people/writers/midgley.html].

102. MISO. *Campaign Against Cruelty: An Animal Activists Handbook*. Leicestershire, England: MISO, 1998. Read the whole book online at [www.veganvillage.co.uk/miso/ handbook/index.htm].

103. Moretti, Laura. *The Good Fight: Speaking for Those Who Can't*. The Best of the Animals' Voice Magazine series. Chico, CA: MBK Publishing, 82

pp., 1994. (1884873243) 25 heartfelt essays on many animal rights topics from 1987-1993. Part of a growing series of excerpts from the *Animals' Voice* magazine. Read articles by the author at [factoryfarming. nettinker.com/farmsanctuary/newsletter/news_moretti1.htm] and [...news_ moretti3.htm].

104. Morris, David B. *Earth Warrior: Overboard with Paul Watson and the Sea Shepherd Conservation Society.* Golden, CO: Fulcrum, 209 pp., 1995. (1555912036) Tales and biography of the radical boat rammer. See also Watson, Paul. Watson characterizes himself as an "environmental Robin Hood" on a campaign to disrupt whaling, sealing, and drift-netting in the world's oceans. See a list of direct actions by Paul Watson at [www. seashepherd.org/actions/acwhale.html]. Is it wrong to enforce rulings by a world oversight commission, like the International Whaling Commission? If the IWC bans whaling of young whales, and yet is incapable of enforcing such rulings, is it okay for groups like the Sea Shepherd Society to enforce them? To some degree, this is what they claim to be doing. Who, if anyone, should be enforcing such rules? Right now, no government is enforcing them.

105. Newkirk, Ingrid. *Free the Animals: The Untold Story of the Animal Liberation Front and Its Founder "Valerie."* New York: Noble, 372 pp., 1992. (187936011x) The head of PeTA defends the Animal Liberation Front. The Animal Liberation Front is known for its "direct actions" of sabotage and property damage to stop laboratory experimentation, fur ranching, and other such industries. Gives a real human face to these "terrorists" who are being pursued by the FBI; they are fairly normal people acting in unusual ways based on their conviction that laws may be broken to save animals. Several direct actions are recounted. See an Animal Liberation Front Frequently-Asked-Questions page at [www. hedweb.com/alffaq.htm] and a related article at [www.csmonitor.com/ durable/1997/08/29/us/us.7.html].

106. Newkirk, Ingrid. *You Can Save the Animals: 251 Ways to Stop Thoughtless Cruelty.* Rocklin, CA: Prima, 288 pp., 1999. (0761516735) The President of PeTA (People for the Ethical Treatment of Animals) on legal ways to save beasts, in our role as consumers, letter-writers, pet adopters, etc. Stop your own practices of hunting and fishing, refuse to do dissection in the classrooms, and stop eating meat. Stopping vivisection, meat-eating, entertainment, animal sports, even pets? See the organization's homepage at [www.peta-online.org]. The author is a real firebrand to the anti-animal rights movement, as she frequently says controversial things, like "a pig is a rat is a dog is a boy." PeTA is the group that opposes animals as pets,

which removes them from the central constituency of animal rights believers, who tend to be pet-owning and pet-loving single women.

107. Nordquist, Joan, compiler. *Animal Rights: A Bibliography.* Contemporary Social Issues #21. Santa Cruz, CA: Reference and Research Services, 68 pp., 1991. (0937855405)

108. Noske, Barbara. *Beyond Boundaries: Humans and Animals.* Montreal: Black Rose, 253 pp., 1997. (1551640783) "The first part focuses particularly on the phenomenon of human domestication of animals, a (forced) human-animal relationship which shows poignantly the extent to which humans have come to dominate and exploit animals as mere resources. . . . The latter part attempts to look at the human-animal relationship . . . from the animal's standpoint, and to find ways through which animals can have their integrity and subjectivity restored to them while at the same time their Otherness is respected."(p. ix) A study of human and animal relationships for sociology and anthropology students. This is a moderate feminist view. Read a brief article by the author at [www.rnw.nl/racism/noskeen.html]; see the table of contents and a brief biography at [www.web.net/~blakrose/beyondb.htm].

109. Oliver, Daniel T. *Animal Rights: The Inhuman Crusade*, 2nd ed. Studies in Organizational Trends, volume 13. Merril Press, 260 pp., 1993. (09367-83230) Read a major article by the author at [www.capitalresearch.org/ap/ap-0898.html].

110. Orlans, F. Barbara, et al. *The Human Use of Animals: Case Studies in Ethical Choice*. New York: Oxford University, 352 pp., 1997. (01951-19088) How much can we harm animals in order to benefit human society while still being ethical? Sixteen essays. Reviewed at [www.upc-online.org/spring99/review_hua.html]. See the table of contents at [www.oup-usa.org/toc/tc_0195119088.html].

111. Paterson, David, and Palmer, Mary, eds. *The Status of Animals: Ethics, Education, and Welfare*. Wallingford, England: CAB International, 268 pp., 1989. (0851986501)

112. Perez, Berta Elena. *Ideology of the Animal Rights Movement.* Ph.D. dissertation, University of Minnesota. Ann Arbor, MI: University Microfilms, 1990. I have not been able to review this document.

113. Petrinovich, Lewis F. *Darwinian Dominion: Animal Welfare and Human Interests.* Cambridge, MA: MIT, 448 pp., 1998. (0262161788) The author is a psychology professor at University of California, Riverside. He says

that humans are unique and animals can rightly be used by humans because of their "lower" natures. This could be taken as a political application of Darwinism known as social Darwinism; that the survival-of-the-fittest concept may justify any actions by domination of one species over another. I doubt that Darwin intended for his biological theories to become justification for political proclamations, as the Nazis also seemed to be attempting in their eugenics and "purification" of the human species.

114. Pluhar, Evelyn B. *Beyond Prejudice: The Moral Significance of Human and Non-Human Animals.* Durham, NC: Duke University, 370 pp., 1995. (082231648x) The author, a philosophy professor at Penn State University, Fayette, presents the view that speciesism (belief that humans are more important than animals) is a form of bigotry. Sentience should be the main criteria for moral consideration. Reviewed at [www.montelis.com/satya/backissues/jan97/prejudice.html]. See the author's homepage at [www.personal.psu.edu/faculty/e/x/exp5/].

115. Preece, Rod. *Animals and Nature: Cultural Myths, Cultural Realities.* Seattle: University of Washington Press, 336 pp., 1999. (0774807245) The author is a political science professor at Wilfrid Laurier University, influenced by Stephen R.L. Clark. "The theme of this book is that the West has been significantly more sympathetic to the natural realm than it has been given credit for. By contrast, the customary depiction of Oriental and Aboriginal concerns for the natural realm, it is argued, has been greatly overdrawn." (p. xi) Read an excerpt and the table of contents at [www.ubcpress.ubc.ca/featured/anim-e.htm].

116. Preece, Rod, and Chamberlain, Lorna. *Animal Welfare and Human Values.* Waterloo, Ontario: Wilfred Laurier University, 334 pp., 1993. (08892-02273) The authors are leaders of an SPCA (Society for the Prevention of Cruelty to Animals) chapter in Canada. They propose using "primal sympathy" (from the poet William Wordsworth) and a sense of universal "community" to solve our problems of harming animals for the purposes of vivisection, fur, and meat. Chapters include discussions on experimentation, fur, hunting, factory farms, entertainment, and philosophy. The authors also attack "the arrogant fanaticism of some of the animal liberationists who imagine there are simple moral solutions to complex social problems." (p. 1) The authors sometimes side with animal rights, and sometimes against. Briefly reviewed at [www.geocities.com/Athens/7259/review.html].

117. Price, Jennifer. *Flight Maps: Adventures with Nature in Modern America.* Basic Books, 325 pp., 1999. (0465023858) A scholarly work with good notes and illustrations, looking for explanations as to human dealings with

animals and nature. Includes studies of the extinction of bird species for women's hat adornments.

118.Pringle, Laurence. *The Animal Rights Controversy.* New York: Harcourt Brace Jovanovich, 112 pp., 1989. (0152035591) Written for young adults.

119.Ray, Dixy Lee, and Guzzo, Lou. *Environmental Overkill: Whatever Happened to Common Sense?* Washington, DC: Regnery Gateway, 260 pp., 1993. (0895265125) The former governor of Washington State (Ray) counters "leftist" statistics on the environment and animal rights with counter statistics and arguments. One chapter is on endangered species. The author is very critical of the Endangered Species Act. Reviewed at [www.u-turn.net/1-3/overkill.html]. One of the dangers to animal rights groups is the use of exaggerated statistics in trying to increase funding. This simply makes an opening for their opponents to discredit their information, which results in the public becoming suspicious of all their statistics (the old "Chicken Little" syndrome — the sky is not falling). I suspect that while these authors go too far in their attacks, they likely have correctly debunked some of the inaccurate information that is frequently disseminated in the unending search for soundbites and video clips.

120.Ray, P. M. *The Teaching of Animal Welfare* (in two parts). London: Universities Federation for Animal Welfare, 1988. (0900767537 and 0900767545) These were used for classes in veterinary science and agricultural science.

121.Reece, Kathleen A. *Animal Organizations and Services Directory, 1992-93,* 4th ed. Huntingdon Beach, CA: Animal Stories, 170 pp., 1993. (0961620226) See links to many animal rights groups at [www.cnw.com/~maser/backyard/Animal.html].

122.Regan, Tom. *All that Dwell Therein: Animal Rights and Environmental Ethics.* Berkeley: University of California, 249 pp., 1982. (0520045718) Ten essays and lectures on vegetarianism, whaling, native Americans, experimentation, rights, and critique of the use of utilitarian philosophy by Peter Singer. Read a brief article by the author at [arrs.envirolink.org/ar-voices/regan_deep.html].

123.Regan, Tom. *The Case for Animal Rights.* Berkeley: University of California, 422 pp., 1985 (originally published 1983). (0520054601) The author uses "moral rights theory" to mount a defense of moral rights for animals. He says that animals have "equal inherent value." This work is scholarly and philosophical, with complex issues and complex answers. David Degrazia wrote that this book is "perhaps the most systematic and explicitly worked-out book in animal ethics . . . carefully argued and thorough." Read an article by the author at [arrs.envirolink.org/ar-voices/case_for_ar.html] and a biography at [www.ivu.org/people/writers/regan.html].

124. Regan, Tom, and Singer, Peter, eds. *Animal Rights and Human Obligations.* Englewood Cliffs, NJ: Prentice Hall, 250 pp., 1989 (originally published in 1976). (0130368644) Two leading animal rights philosophers present an overview of historical and modern writings on the nature and treatment of animals. "Although human beings eat other animals, experiment upon them, and destroy their habitats, we rarely pause to consider whether our practices toward them are ethically defensible." Read some quotations from the book at [www.xenos.org/ministries/c&c/c_canim.htm].

125. Reichmann, James B. *Evolution, Animal 'Rights', and the Environment.* Washington, DC: Catholic University of America Press, 400 pp., 2000. (0813209544) Not yet published. Read a description of the upcoming work on the [www.amazon.com] website. The author teaches philosophy at Seattle University. This work discredits animal rights philosophy (and vegetarianism) using Darwinianism and natural law, including specific references to the works of Tom Regan and Peter Singer. See the author's bibliography at [www.seattle.edu/artsci/departments/philosophy/philfac.htm].

126. Rich, Elizabeth, ed. *National Guide to Funding for the Environment and Animal Welfare*, 4th ed. New York: Foundation Center, 527 pp., 1998. (0879547707) This is one of a series of books that provide state-by-state listings of charitable organizations offering funds to groups involved in specific actions, like animal rights. Groups use this guide to write grant requests for funding. Each entry includes contact information, types of grants offered, and application guidelines. See a sample entry at [fdncenter.org/marketplace/catalog/fie.html].

127. Rodd, Rosemary. *Biology, Ethics, and Animals.* Oxford: Clarendon, 272 pp., 1990. (019824052x) Analyzes philosophy and biology to change people's attitudes towards animals and to improve communication between activists and opponents. She puts humans ahead of animals in moral importance but wants improvements made for their welfare. Sometimes the book is hard to read due to a lack of clear organization.

128. Rollin, Bernard E. *Animal Rights and Human Morality*, revised ed. Buffalo, NY: Prometheus, 248 pp., 1992 (originally published in 1981). (0879757892) The first chapter, on moral rights, is an outstanding sum-

mation of various philosophical positions in relation to animal rights. Other topics include legal rights, animal research, and the keeping of pets. Why do scholars try to separate ethics from science? And why do people believe that animals are incapable of feeling pain? The author is a professor at Colorado State University. Download an article by the author at [www.ansc.purdue.edu/wellbeing/FAWB1993/Rollin.pdf].

129. Rose, Tom. *Freeing the Whales: How the Media Created the World's Greatest Non-Event*. New York: Carol, 307 pp., 1989. (1559720115) A 1991 Reader's Digest "Today's Best Non-Fiction" Pick. Read small excerpts at [www.highnorth.no/fr-th-wh.htm]. Though I am sympathetic to the desire to free the two trapped whales from the ice in the Arctic Circle, I must say that the media hype was not helpful overall to the animal rights cause. This is not because the effort to save the whales was ignoble, but because the Russians got a lot of good publicity and credit for saving the whales (by sending a couple of icebreakers), when in fact Russia is (or was) probably the largest poacher of the world's whales, driving them toward extinction. Those couple of days of good media coverage probably reversed a decade worth of negative publicity over the Russians' illegal whaling.

130. Rowe, Martin, ed. *The Way of Compassion: Survival Strategies for a World in Crisis*. New York: Stealth Technologies, 256 pp., 1998. (09664-05609) Discussion of the issue facing the animal rights and vegetarian movements of today. Includes essays by Andrew Linzey, Carol Adams, John Robbins, Jane Goodall, Jim Mason, Roger Fouts, and Henry Spira, among others. See the table of contents at [www.continuum-books.com/rowe.htm]. Reviewed at [www.api4animals.org/Publications/AnimalIssues/1999-Spring/Reviews99-01.htm].

131. Rowlands, Mark. *Animal Rights: A Philosophical Defense*. New York: St. Martin's, 200 pp., 2000. (031221720x) See the planned table of contents on the Amazon bookstore site [Amazon.com].

132. Ryder, Richard Dudley. *Animal Revolution: Changing Attitudes toward Speciesism*. Oxford: Blackwell, 385 pp., 1989. (0631152393) This writer first penned the concept of "speciesism" as bigotry against animals. This work is a historical survey of the relationship between man and animals, with a special emphasis on the British movement since 1969.

133. Ryder, Richard Dudley. *The Political Animal: The Conquest of Speciesism*. Jefferson, NC: McFarland, 147 pp., 1998. (0786405309) The author coins a new term, "painism," saying that we have a moral duty to reduce the pain of all life on Earth. Read a brief article by the author at [www.newscientist.com/nsplus/insight/animal/ryder.html].

134. Salt, Henry S. *Animals' Rights Considered in Relation to Social Progress*. Clarks Summit, PA: Society for Animal Rights, 240 pp., 1980 (originally published 1894). (0960263209) This author was perhaps the greatest of the nineteenth-century animal-issues thinkers, writing several books and pamphlets on the subject. Many of his ideas became pillars of the modern AR movement. He said that animals are individuals, not tools for human use, and that justice would leave them free to develop naturally. He also said that sympathy is the first step in the proper human reaction to animal issues, then a realization of rights will follow. Read excerpts at website [www.envirolink.org/arrs/essays/H_Salt.html#excerpt] and an article about the author at [arrs.envirolink.org/ar-voices/H_Salt.html].

135. Sapontzis, S. F. *Morals, Reason, and Animals*, reissue ed. Philadelphia: Temple University, 302 pp., 1992 (originally published 1987). (08772-29619) Three major goals of our moral tradition are: reducing suffering, being fair, and developing moral virtues. Based on these goals, the author condemns much of our culture's consumption of animals but does not advocate a total ban on animal research. Read an article by the author at [phil.indiana.edu/ejap/1995.spring/sapontzis.1995.spring.html].

136. Scarce, Rik. *Eco-warriors: Understanding the Radical Environmental Movement*. New York: Noble, 291 pp., 1990. (096226833x) The author is a sociology professor in Montana. While writing his Ph.D. on this subject he interviewed many animal rights activists who were wanted by the FBI. He was jailed for several months in 1993 for refusing to give information on the fugitives' whereabouts, claiming scholarly confidentiality (like journalistic confidentiality). Included are histories of the Animal Liberation Front, Earth First!, and the Sea Shepherd Conservation Society. See links to environmental activism websites at [www.worldrevolution.org/ResourcePages/Environment.html].

137. Schweitzer, Albert. *Reverence for Life*, 2nd ed. New York: Irvington, 74 pp., 1993. (0891979204) This is a collection of Schweitzer's ethical statements, put together after his death. Chapter one is titled "Feeling for Animal Life." See more at [www.portalproductions.com/spiritnature/Schweitzer.htm]. Read an excerpt at [www.pcisys.net/~jnf/schauth.html#RQ8]. Read a brief biography of Schweitzer at [www.Schweitzerfellowship.org/91text.html].

138. Screaming Wolf (pseudonym). *A Declaration of War: Killing People to Save Animals and the Environment*. Grass Valley: Patrick Henry Press, 1991. (0962925977) This book created a firestorm of controversy and is almost impossible to locate; it is outlawed in Canada and the United Kingdom for "inciting violence." Sidney and Tanya Singer claim to have

received the manuscript anonymously in the mail, though some believe they wrote it. They are members of an animal rights group that practices direct action tactics. See related websites at [www.animalrights.net/terrorism_report/extent.html] and [www.eskimo.com/~rarnold/DOJReport.html]. The basic belief represented in Screaming Wolf's book is similar to that of extreme anti abortionists: if killing animals is immoral, then it is no crime to use force to stop that killing. Thus far the actual perpetration of such deeds has been rare, aside from the Unabomber (sending mail bombs to industrialists) and a few rare attempts at bombing vivisectionist vehicles. Though many animal rights people do claim to believe that animals and humans are of equal value, there has not yet been a groundswell of support for violence against animal killers.

139. Scruton, Roger. *Animal Rights and Wrongs.* London: Demos, 113 pp., 1996. (1898309825) The author asks how we should think about the morality between animals and people in relationships. He seems to conclude that animals do not have rights. Read an article by the author at [www1.pos.to/~luna/whale/gen_art_scru.html]. Reviewed at [vzone.virgin.net/old.whig/fl28pets.htm].

140. Sherry, Clifford J. *Animal Rights: A Reference Handbook.* Santa Barbara, CA: ABC-CLIO, 214 pp., 1994. (0874367255) A general-purpose work offering history, legislation, arguments pro and con, biographical sketches of activists, and annotated listings of animal rights sources.

141. Shorto, Russell, et al. *Careers for Animal Lovers.* Brookfield, CT: Millbrook Press, 64 pp., 1992. Written for young adults. See a good bibliography at [www.seaworld.org/zoo_careers/zcreferences.html]. See related sites at [www.tamu.edu/ethology/TipsCar.html] and [acunix.wheatonma.edu/kmorgan/AB_Careers/animal_behavior_careers.html].

142. Silverman, B. P. Robert Stephen. *Defending Animals' Rights Is the Right Thing to Do,* reprint ed. New York: Shapolsky, 228 pp., 1994 (originally published in 1991). (0874367336) This is a very strongly worded demand for equal rights of humans and animals. It is quite graphic in language and pulls no punches in indicting animal rights opponents.

143. Silverstein, Helena. *Unleashing Rights: Law, Meaning, and the Animal Rights Movement.* Lansing: University of Michigan, 320 pp., 1996. (0472106856) How animal rights activists use words in legal ways as legal terms to promote their cause. This is a study with relevance for political science, law, and sociology. Reviewed at [www.montelis.com/satya/backissues/may97/editorial.html].

144. Singer, Peter. *Animal Liberation: A New Ethic for Our Treatment of Animals*, 2nd revised ed. New York: New York Review of Books, 320 pp., 1990 (originally published in 1975). (0940322005) He uses utilitarian philosophy to destroy behaviorist presuppositions on the nature of animals. *Newsweek* credits this book with starting the animal rights movement. It discusses the issues of dominion, vegetarianism, vivisection, factory farming, and equality. Reviewed at [asa.calvin.edu/ASA/book_reviews/12-92.htm]. See an anti-Singer article with a rather threatening title, "Dissecting Peter Singer," at [www.naiaonline.org/disectps.html].

145. Singer, Peter. *The Animal Liberation Movement: Its Philosophy, Its Achievements and Its Future*. Nottingham, England: Old Hammond Press, 1986. Read an interview with the author at [www.animalsagenda.org/features/singer1.html (and also) singer2.html]. Find a good links page at [www.computan.com/~klawrenc/articles/ mar96.html]. Singer is controversial, especially for his views that imperfect human babies may ethically be killed up to one month after birth (not a popular view among people with disabilities).

146. Singer, Peter. *Ethics into Action: Henry Spira and the Animal Rights Movement*. Lanham, England: Rowman & Littlefield, 192 pp., 1998. (0847690733) Singer uses the life and work of activist Henry Spira (founder of Animal Rights International) as an example of how other animal rights activists can succeed. He discusses the lessons that Spira learned in doing civil rights work, trade union bargaining, and animal rights work. There are hints for improvement and warnings about the pitfalls of activism. Reviewed at [www.rasheit.org/VY-LIBRARY/libschwartz36.htm#anchor343135] and [www.api4animals.org/Publications/AnimalIssues/1998-Winter/Reviews98-04.htm].

147. Singer, Peter, ed. *In Defense of Animals*. New York: Harper & Row, 224 pp., 1985. (0060970448) A collection of articles and essays that puts the movement into historical and ethical perspective, by outstanding authors like Tom Regan, Marian Dawkins, Stephen Clark, Mary Midgley, Richard Ryder, Jim Mason, Lewis Regenstein, and Henry Spira. Read an article by the author at [arrs.envirolink.org/ar-voices/pain.html]. Read a two part negative article, largely about Singer's ideas, at [www-bioc.rice.edu/~csmiller/greenazi/vegan.html].

148. Singer, Peter. *The Expanding Circle: Ethics and Sociobiology*. Oxford: Clarendon, 190 pp., 1986 (originally published in 1981). (0192830384) The author uses utilitarian philosophy to promote an animal rights agenda. Read an interview with the author at [cogweb.english.ucsb.edu/Debate/

SingerPM.html]. See a brief biography and excerpts at [www.nettinker.com/ivu/people/writers/psinger.html].

149. Sperling, Susan. *Animal Liberators: Research and Morality*. Berkeley: University of California, 247 pp., 1988. (0520061985) The author, an evolutionary anthropologist, talks to nine activists. See a brief publisher's review at [www.ucpress.edu/books/pages/2334.html].

150. Spiegel, Marjorie. *The Dreaded Comparison: Human and Animal Slavery*, revised and expanded ed. New York: Mirror, 108 pp., 1997 (originally published in 1988). (0962449342) The author founded The Institute for the Development of Earth Awareness. This work uses a lot of quotations. Reviewers are split: some loving the book, some hating it. Read two reviews at [www.montelis.com/satya/backissues/march97/comparison.html] and [www.nsplus.com/nsplus/in sight/animal/respect.html].

151. Spira, Henry. *Strategies for Activists: From the Campaign Files of Henry Spira*. New York: Animal Rights International, 1989. This man has done activist work for trade unions, civil rights, and animal rights. His advice on effective activism should be noticed by those who seek success in their own endeavors. Read an article by Spira at [www.montelis.com/satya/back issues/feb97/activist.html].

152. Steeves, H. Peter, ed. *Animal Others: On Ethics, Ontology, and Animal Life*. Contemporary Continental Philosophy Series. Albany: State University of New York, 256 pp., 1999. (0791443094) "Animal Others brings together original contributions that explore the status of animals from the continental philosophy perspective . . . in the work of philosophers such as Husserl, Heidigger, Nietzche, Merleau-Ponty, and Derrida." (back cover) The essay by Steeves is very odd yet very interesting: including theories on zoos and pets, among other things. European philosophies about the natures, moral status, and roles of animals in society. Brief publishers review at [www.sunypress.edu/sunyp/ backads/html/c43094. html].

153. Stone, Christopher D. *Should Trees Have Standing? Toward Legal Rights for Natural Objects*, reprint ed. New York: Avon, 102 pp., 1996 (originally published in 1972). (0379213818) This work shows how persons who were once not considered to be persons in the eyes of the law (i.e., women, minorities) gained legal rights, and how activists might use these tactics to extend legal rights to natural rivers, forests, and so on. The implications of this are quite broad. See the table of contents and a brief review at [www.oceanalaw.com/books/n156.htm].

154. Strand, Rod, and Strand, Patti. *The Hijacking of the Humane Movement: Animal Extremism*. Wilsonville, OR: Doral, 174 pp., 1993. (0944875289) This is another "love it or hate it" work; hated by animal rights groups, and loved by their opponents. The authors call animal rights activists "a hate group" and "terrorists." The authors are dog breeders, and founded the National Animal Interest Alliance (anti-animal rights group). "[T]hrough the use of misinformation, public humiliation, intimidation and terrorism, the practice of animal rights hurts people." Reviewed at [www.naiaonline.org/hijack.html]. See a chronology of recent animal rights direct actions at [www.naiaonline.org/arterror.html].

155. Styles, John. *The Animal Creation: Its Claims on Our Humanity Stated and Enforced*, reprint ed. Mellen Animal Rights Library #H1. Lewiston, NY: Mellen, 248 pp., 1996 (originally published in 1839). (0773487107) One of the earliest animal rights works, this is a catalog of abuses and cruelty. The author attacks cruelty using science and theology. "No being can be so dissimilar as Jesus Christ and a creature whose bosom is the seat of cruelty." (p. iv)

156. Sweeney, Noel. *Animals and Cruelty and Law*. Bristol, England: Alibi, 119 pp., 1990. (1872724000) A practicing barrister (British lawyer) argues for animal rights from a legal standpoint.

157. Tester, Keith. *Animals and Society: The Humanity of Animal Rights*. New York: Routledge, 218 pp., 1991. (0415047323) A critique of the modern animal rights movement and its philosophies. A history of social and cultural definitions of "animals" and the subsequent believed similarities and differences in comparison to humankind. Tester says that animal rights philosophers should "realise that they are fetishistically upholding obligations which are made and not found."

158. Tobias, Michael. *Voices From the Underground: For the Love of Animals*. Pasadena, CA: New Paradigm Books, 166 pp., 1999. (0932727484) A book of personal accounts of people who work in animal rights and animal welfare. See several short reviews on the [www.amazon.com] website.

159. Turner, E. S. *All Heaven in a Rage: A History of the Prevention of Cruelty to Animals in Great Britain*, reprint ed. New York: Centaur, 324 pp., 1992 (originally published in 1964). The title is based on a poem by William Blake called "Auguries of Innocence," which you can read online at [www.geocities.com/Bourbon Street/Delta/5718/blake.htm].

160. Watson, Paul. *Sea Shepherd: My Fight for Whales and Seals*. New York: Norton, 258 pp., 1982. (0393014991) The author muses on his practices of

throwing dye on pelts, ramming whaling boats, and other radical actions. Very interesting early history of Greenpeace (before Watson was thrown out) and the formation of the Sea Shepherd Society. Throws a bad light on Canadian politics, which protects exploitative industries. Read an article by the author at [arrs.envirolink.org/ar-voices/politics_of_ extinction.html].

161. Watson, Paul, and Mowat, Farley. *Ocean Warrior: My Battle to End the Illegal Slaughter on the High Seas*. Toronto: Key Porter, 264 pp., 1995. (1550135694) A fascinating look at a very active form of opposition to whaling, sealing, and driftnetting. Watson is stubborn and strong-willed, seeing himself as a pirate against sea exploiters. He holds strong grudges against Greenpeace and the Canadian government, but perhaps with good reason. His actions have a great impact in drawing media attention and forcing government action on violation of sea laws. See the author's group, The Sea Shepherd Conservation Society, at [www.seashepherd.org]. He alienates some animal rights people because he is not a vegetarian, in spite of his more active AR tactics with seals and whales. Read an article by the author at [www.seashepherd.org/essays/caplog962.html].

162. Webster, John. *Animal Welfare: A Cool Eye Towards Eden*. Oxford: Blackwell, 273 pp., 1995. (0632039280)

163. Wilkins, David B. *Animal Welfare Legislation in Europe: European Legislation and Concerns*. The Hague: Kluwer, 448 pp., 1997. To see the table of contents and a brief publisher review, go to [www.kluwer.com] and do a search for Wilkins.

164. Williams, Jeanne, ed. *Animal Rights and Welfare*. Reference Shelf vol. 63 no. 4. New York: H. W. Wilson, 168 pp., 1991. (0824208153) Seventeen essays by various authors including Jane Goodall. Eight chapters on vivisection, others on fur, hunting, and food animals.

165. Willis, R. G., ed. *Signifying Animals: Human Meaning in the Natural World*. One World Archaeology series. London: Routledge, 288 pp., 1994. (0415095557) This book seems to be an analysis of animal symbolism in human cultures. Why do we see snakes as evil? Why are doves seen to be pure?

166. Wise, Steven M. *Rattling the Cage: Toward Legal Rights for Animals*. Cambridge: Perseus Books, 362 pp., 2000. (0738200654) Wise is a lawyer who teaches animal rights law at several schools, including Harvard University. He proposes that there is now enough evidence to support legal rights for at least the "higher" primates: specifically, chimpanzees and bonobos. See a brief description at [www.2think.org/rattling.shtml]. Jane

Goodall wrote the forward. The book is receiving good reviews; see a recent one at [www.nytimes.com/books/00/02/20/reviews/000220.20sun stet.html]. Read an article about legal rights for animals at [www.lclark. edu/~alj/intro3. Html].

167. World Animal Net. *World Animal Directory*. Boston: World Animal Net, 305 pp., 1999. 10,000 addresses of groups in 130 countries, including website addresses. You can use it online at [worldanimal.net/ wan.htm].

168. Wynne-Tyson, Jon, ed. *The Extended Circle: a Commonplace Book of Animal Rights*. New York: Paragon House, 619 pp., 1989. (1557781486) A collection of quoted passages from thinkers of all ages that are relevant to animal rights. Read an article by the author at [www.viva.org.uk/ Viva!Guides/future.html].

Chapter 2

Animal Natures

This chapter contains most of the scientific evidence behind animal rights arguments regarding the question, "What is an animal?" The answer to this question is the basis for any rational position on animal rights issues, aside from possible religious perspectives. Here are the types of issues to be found in the works listed in this chapter.

1. Are animals alive? How is this different from plant life?
2. Are animals conscious? Are they sentient?
3. What is instinct? Does instinct explain the whole panorama of observed animal behaviors?
4. What are valid observations in the study of animal behavior (ethology)? Can valid observations be made in captivity, or must the creatures be observed in their wild habitats?
5. Are animals intelligent? Can they recognize their kin? Can they recognize themselves in a mirror?
6. Does animal communication equal language?
7. Do animals have emotions and feelings? Is their psychology at all similar to human psychology? Why do animals play?
8. Are animals social creatures? Do their societies resemble human groups? Is group cooperation the same as human altruism and love?
9. Do animals have any sense of morality or conscience? Are animal deceptions really the same as human lies? Religion? Do they create art, or music?
10. Do animals feel pain, or are their screams just "instinctual vocalizations?"
11. What is anthropomorphism? Are most ethologists and animal rights people simply reading human values into observed animal behaviors?

The study of animal natures has been going on for millennia, at least as far back as ancient Greece. However, in most cases these studies had a strictly religious foundation. How did the God(s) make these creatures? What do the God(s) want us to do with the animals? The answers to these questions were generally made from the prevalent religious views of the day. Although the role of religions has been widely maligned by the animal rights movement, and some religious perspectives have been correctly accused in this regard, it was not until the advent of anti religion that animal abuse became widespread. While religious folk might not have considered animals to be of great importance, the rise of "science-as-god" sent religion and ethics out the window together. Rationalism and the new scientific method encouraged the dissection and analysis of all living things, and had no religious or ethical compunctions to live by. Descartes declared that animals are mindless automata, mere biological robots, whose screams under the vivisector's knife were simply instinctual reactions to stimuli, not a sound of "real" pain.

The work of Charles Darwin, while feared by religious folk as a death-knell to belief in God as creator, did not support that of the anti-ethical ideas of Cartesian science. Though he did promote the view that man was descended from lower biological forms, counter to religious claims of creaturehood, he said that animals are very much like people, including their physiological, psychological, and emotional states. In other words, if man is descended from animals, then animals cannot be all that different from us, and we ought to protect and respect their likeness to us. The followers of Descartes (known as Cartesians) continued to reign supreme in the halls of science, but Darwin's ideas grew and spread throughout the early twentieth century and have become more prominent in modern times. It would be no stretch to say that animal rights owes a great deal to Charles Darwin, and that it could not have spread so effectively without the scientific studies of ethologists (students of animal behavior) after him.

Probably the earliest ethologists to gain popular notice (and a Nobel Prize) were Konrad Lorenz, Nikolaas Tinbergen, and Karl Von Frisch, who studied bees and birds, and discovered that "instinct" might not be the end-all-be-all of animal existence that Cartesians claimed. Instinct might not explain the variety of animal behaviors that require flexibility and alteration in activities.

Some animal rights proponents have not gone this scientific route in the search for credibility and societal changes. Many have taken the religious route, promoting Eastern philosophies, pantheism, and Gaia theory (often mixed with feminism), saying that all life is equal and must be accepted as equal regardless of the scientific findings.

There is also an intense mistrust between the opposing camps. Many scientists will simply ignore the animal rights proponents as "wackos," while the animal rights people shout "murder" at those who perform animal experiments. The atmosphere of hostility is not conducive to debate or compromise. True, many people involved on both sides do not want any compromise. It has become an all-or-nothing proposition, without room for discussion. However, there are many who rest in more moderate positions and do seek a dialogue on these issues. For those people, these resources regarding the scientific evidence of animal natures will be of help.

---

169. Adler, Mortimer J. *The Difference of Man and the Difference it Makes.* New York: Fordham University, 395 pp., 1993 (originally published in 1967). (0823215350) The founder of the Institute for Philosophical Research seeks to prove vital differences between man and animals in intelligence and self-awareness. This work is based on lectures he gave at the Encyclopedia Brittanica lectures in 1966. For more on his ideas regarding animal awareness, see the website [www.harborside.com/home/r/radix/adlerman1.htm].

170. Alcock, John. *Animal Behavior: An Evolutionary Approach*, 6th revised ed. Sunderland, MA: Sinauer, 640 pp., 1998 (originally published in 1974). (0878930094) Detailed studies for college students doing evolutionary behavioral research. Written (with some light humor) by a biology professor from Arizona State University. This is a textbook at many schools. See a brief biography at [ls.la.asu.edu/biology/faculty/alcock.html]. Read a publisher review and table of contents at [www.webcom.com/~sinauer/saw/Titles/Text/alcock. html].

171. Allen, Colin, and Bekoff, Marc. *Species of Mind: The Philosophy and Biology of Cognitive Ethology.* Cambridge, MA: MIT, 280 pp., 1997. (0262011638) A philosopher (Allen) and cognitive ethologist team up to discuss animal behavior. Read the preface, table of contents, and chapter one at Allen's webpage [logic.tamu.edu/~colin/SpeciesofMind/]. Reviewed at [psyche.cs.monash.edu.au/v4/psyche-4-21-robinson.html], and [www.ex.ac.uk/~WHDittri/anbeh98.html].

172. Antinucci, Francesco, ed. *Cognitive Structure and Development in Non-human Primates.* Comparative Cognition and Neuroscience series. Hillsdale, NJ: Lawrence Erlbaum, 386 pp., 1989. (0805805443) An Italian expert explains mental processes (using Piagetian-style testing) in infants and in evolutionary history. See the publisher review and table of contents at [www.erlbaum.com/118.htm].

173. Association of Veterinary Teachers and Research Workers. *Guidelines for the Recognition and Assessment of Pain in Animals.* London: Universities Federation for Animal Welfare, 1989. (0785537619) See an article on this subject at [www.nsplus.com/nsplus/insight/animal/pain.html]. The popular

view among vivisectionists since Descartes has been that animals cannot feel pain, that they are simply complex biological machines that instinctually vocalize (screech) when touched. Since the animals have no "real mind" they can feel no "real pain." In recent decades this view has been somewhat discredited, and many experiments now require anesthesia or pain-killing medication for the animals.

174. Attenborough, David. *The Trials of Life: A Natural History of Animal Behavior*. Boston: Little, Brown, 320 pp., 1990. (0316057517) Not an ethology textbook, but a popular portrayal of the variety of animal behavior in our world. It was made into a lengthy and famous television series and is available on home video. Each video is briefly reviewed at [www.browbeat.com/browbeat01/snuffilm.htm]. See a good ethology site at [www.nua-tech.com/paddy/ethology.shtml].

175. Balda, Russell P., et al. *Animal Cognition in Nature: The Convergence of Psychology and Biology in Laboratory and Field*. New York: Academic, 480 pp., 1998. (012077030x) Biography on the author at [www.life.uiuc.edu/alumni/russellp.htm]. Read an article about animal intelligence at [www.newscientist.com/nsplus/insight/big3/conscious/day3b.html].

176. Barber, Theodore Xenophon. *The Human Nature of Birds: a Scientific Discovery with Startling Implications*, reprint ed. New York: Penguin, 226 pp., 1994. (0140234942) The author says that birds are comparable to (or greater than) humans in intelligence. He says that birds have abstract thoughts and emotions, and make tools for use. He worked for 30 years as a behavioral scientist and six years in avian studies, and was influenced greatly by Donald Griffin. Read an article by the author about this book at [www.psyeta.org/hia/vol8/barber.htm] and an excerpt at [www.gaialogic.org/gaianation/animalia.html].

177. Bavidge, Michael, and Ground, Ian. *Can We Understand Animal Minds?* New York: St. Martin's, 176 pp., 1994. (0312124244) An investigation of different theories of how minds work. The authors say that the modern scientific understanding of "mind" is seriously flawed in its "reluctance to attribute psychological states and capacities to non-human animals." Read an article by Bavidge at [calliope.jhu.edu/demo/philosophy_psychiatry_and_psychology/3.1bavidge.html].

178. Beamish, Peter. *Dancing with Whales*. St. Johns, Nova Scotia: Creative Publishers, 164 pp., 1993. (1895387280) See a site about Beamish's work at [www.animalcontact.com/research/index.htm]. Reviewed at [www.physics.helsinki.fi/whale/intersp/pages/book1.htm].

179. Bekoff, Marc, and Byers, John A., eds. *Animal Play: Evolutionary, Comparative and Ecological Perspectives*. Cambridge: Cambridge University, 290 pp., 1998. (0521583837) Why do animals play? Based on a 1996 symposium of the Animal Behavior Society, with 14 contributors. See the table of contents and abstracts of the articles at [cogweb.english.ucsb.edu/Abstracts/Bekoff&Byers_98.html].

180. Bekoff, Marc, and Jamieson, Dale, eds. *Interpretations and Explanation in the Study of Animal Behavior: Explanation, Evolution and Adaptation*, 2 volumes. Westview Special Studies series. Boulder, CO: Westview, 970 pp., 1990. (0813377048 and 0813379792) See a bibliography at [mitpress.mit.edu/MITECS/Abstracts/bekoff2_r.html] and a good article about the significance of animal behavior at [www.cisab.indiana.edu/ABS/valueofanimalbehavior.html].

181. Bekoff, Marc, and Jamieson, Dale, eds. *Readings in Animal Cognition*. Cambridge: MIT, 496 pp., 1995. (026252208x) Twenty-four readings, largely a revision of *Interpretations and Explanation in the Study of Animal Behavior* from 1990. Read a publisher review and table of contents at [mitpress.mit.edu/book-home.tcl?isbn=026252208x].

182. Bonner, John Tyler. *The Evolution of Culture in Animals*, reprint ed. Princeton, NJ: Princeton University, 216 pp., 1989 (originally published in 1980). (0691023735) Now considered to be a classic in sociobiology. How do species of animals communicate rules of behavior between their generations? Why and how did culture evolve? The author is a developmental biologist from Princeton University. See a brief author bibliography at [www.santafe.edu/~shalizi/notebooks/bonner.html]. Great links on the subject at [www.humaneevolution.net/tablecontents2.html].

183. Boysen, Sarah T., and Capaldi, E. John, eds. *The Development of Numerical Competence: Animal and Human Models*. Comparative Cognition and Neuroscience series. Hillsdale, NJ: Lawrence Erlbaum, 296 pp., 1993. (0805807497) Boysen is the editor for *The Journal of Comparative Psychology*. Read an interview with Boysen at [www.girlscientist.org/boysen_profile.html]. See a related article at [abcnews.go.com/sections/science/DailyNews/monkeys981022.html].

184. Bradbury, Jack W., and Vehrencamp, Sandra Lee. *Principles of Animal Communication*. Sunderland, MA: Sinauer, 882 pp., 1998. (0878931007) A scholarly work on the physiological and social effects of signals between animals. See the publisher review and table of contents at [www.webcom.com/~sinauer/saw/Titles/Text/bradbury.html#Principles]. See a

biography of Bradbury at [www-biology.ucsd.edu/shadow/sa/newbrochure/bradbury.html].

185.Bradshaw, John L., and Rogers, Lesley J. *The Evolution of Lateral Asymmetries, Language, Tool Use and Intellect*. New York: Academic, 463 pp., 1993. (0121245608) Neuropsychology of animals and humans compared and contrasted. Read related articles at [student-www. uchicago.edu/~mwruan/hemisphericspecialization.htm] and [www.neoteny.org/a/lateralization2.html].

186.Bright, Michael. *Intelligence in Animals*. The Earth, Its Wonders, Its Secrets series. New York: Reader's Digest Association Limited, 160 pp., 1997 (originally published in 1994). (0895779137) Wonderful summary of the wide variety of studies on non-human intelligence, with many photographs and charts. Includes chapters on animal communication, domestication, survival strategies, and cooperation in societies. Read an interesting article about the differences between instinct and intelligence at [acidmagic.com/books/universe-06.html].

187.Broom, Don M., et al. *Stress and Animal Welfare*. London: Chapman & Hall, 211 pp., 1994. (0412395800) The director of veterinary medicine at Cambridge University explains how researchers can use "stress indicators" for measurable assessments of animal condition. See a brief article by the author at [ely.anglican.org/parishes/camgsm/Majestas/1999/April.html]. Read a related article at [www.nsplus.com/nsplus/insight/animal/stressful.html].

188.Budiansky, Stephen. *If a Lion Could Talk: Animal Intelligence and the Evolution of Consciousness*. New York: Free Press, 288 pp., 1998. (0684837102) The author says that science is becoming enamoured with anthropomorphic theories of animal intelligence. He says that the difference between humans and animals is one of kind, not just degree. Read an interview with the author at [www2.theatlantic.com/atlantic/unbound/bookauth/ba981209.htm]. Reviewed at [www.montelis.com/satya/dumb_chums.html] and [cas.bellarmine.edu/tietjen/images/only_unthinking_intelligence.htm].

189.Burton, Frances D., ed. *Social Processes and Mental Abilities in Non Human Primates: Evidences from Longitudinal Field Studies*. Lewiston, NY: Mellen, 283 pp., 1992. (0773495371) The author is an anthropology teacher; see one of her class syllabi on primates at [night.primate.wisc.edu/pin/syllabi/sylburton4]. Reviewed at [www.chass.utoronto.ca/epc/srb/srb/sociality.html].

190. Busch, Heather. *Why Cats Paint: A Theory of Feline Aesthetics*. Berkeley: Ten Speed, 96 pp., 1994. (0898156122) Some say that this work is actually a clever joke, not meant to be taken seriously. However, many readers do take it seriously. Since some animals, like elephants, do seem to have an artistic interest (see David Gucwa and James Ehmann and Dick George in this chapter), maybe this is serious. Read a related article at [www.illinois times.com/1998/feb0598/news.html]. See cute feline-art sites at [www. mania.com.au/~pshaw/art.html] and [www.netlink.co.nz/~monpa/].

191. Byrne, Richard W. *The Thinking Ape: Evolutionary Origins of Intelligence*. New York: Oxford, 280 pp., 1995. (0198522657) Chapter eleven is on ape experiments in language. Read an article by the author at [www.cogsci.soton.ac.uk/bbs/Archive/bbs.byrne.html].

192. Candland, Douglas Keith. *Feral Children and Clever Animals: Reflections on Human Nature*. New York: Oxford University, 411 pp., 1993. (01950-74688) The author is a psychology professor at Bucknell University. An analysis of how scientists and people view stories of comparative psychology, like stories about chimps that write, the seemingly mathematical horse Clever Hans and wolf children. We usually end up making conclusions that exactly parallel what we believed beforehand. Reviewed at [www.psyeta.org/sa/sa3.2/jasper.html]. See an article about the phenomenon of feral children at [www.halcyon.com/badams/feral. htm].

193. Cavalieri, Paola. *The Great Ape Project: Equality Beyond Humanity*. New York: St. Martin's, 312 pp., 1994. (031211818x) Thirty-four essays arguing that apes are close cousins to humans. Peter Singer, Marian Stamp Dawkins, and Jane Goodall are among the writers. Reviewed at [www. montelis.com/satya/backissues/dec96/bookreview1.html]. See also [www. nsplus.com/nsplus/insight/animal/freeapes.html].

194. Cheney, Dorothy L., and Seyfarth, Robert M. *How Monkeys See the World: Inside the Mind of Another Species*, reprint ed. Chicago: University of Chicago, 378 pp., 1992 (originally published in 1990). (0226102467) A biologist and psychologist from the University of Pennsylvania examine primate deception, intelligence, communication, and behavior. They spent much of their time studying kin recognition and vocalizations among Vervet monkeys. Reviewed at [www.anatomy.usyd.edu.au/danny/book-reviews/h/How_Monkeys_See_the_World.html].

195. Colgan, Patrick. *Animal Motivation*. New York: Routledge, 159 pp., 1989. (0412318504) A scholarly work for zoology and psychology students. Why

do animals act in certain ways? See a site on animal motivation at [www2.abtech.net/%7Etni/Motives.html].

196. Corballis, Michael C. *The Lopsided Ape: The Evolution of the Generative Mind.* New York: Oxford, 366 pp., 1991. (0195083520) An interesting theory, that human and primate brains developed differently during the eons of evolution because language is a creative function of the left brain. Humans tend to be right-handed, and therefore left-brained, while primates are ambidexterous. Read an article by the author at [www.pdc.co.il/corballi.htm]. Reviewed at [www1.dragonet.es/users/markbcki/corbal.htm].

197. Coren, Stanley. *The Intelligence of Dogs: A Guide to the Thoughts, Emotions, and Inner Lives of Our Canine Companions.* New York: Bantam, 271 pp., 1995. (0553374524) The author is a psychology professor at the University of British Columbia. This book caused widespread popular controversy by presenting evidence that different breeds of dogs have greatly varied degrees of intelligence. Many pet owners were offended that their own dog might be less intelligent than the neighbor's dog. See his top 100 canine brain breeds at [www.petrix.com/dogint/intelligence.html]. See also the website [www.animalnews.com/coren].

198. Crail, Ted. *Apetalk and Whalespeak: The Quest for Interspecies Communication.* Los Angeles: Tarcher, 298 pp., 1981. (0809255278) Considers the questions of animal consciousness and how animals can communicate amongst themselves and with people. Interviews with leading researchers on gorillas, dolphins, and whales. Interesting notes on the political ramifications of such research: how universities often stonewall and force repudiations when conflicting with "accepted dogma." Read a related essay at [www.dolphininstitute.org/isc/text/essay8.htm].

199. Crist, Eileen. *Images of Animals: Anthropomorphism and Animal Mind.* Animals, Culture, and Society series. Philadelphia: Temple University, 256 pp., 1999. (1566396565) The author is critical of new ideas about "animal minds." She says that such anthropomorphic tags will burden our society and endanger civilization. Read a publisher review at [www.temple.edu/tempress/titles/1257_reg.htm].

200. Dachowski, Lawrence W., and Flaherty, Charles F., eds. *Current Topics in Animal Learning: Brain, Emotion, and Cognition.* Hillsdale, NJ: Lawrence Erlbaum, 437 pp., 1991. (0805804412) Thirteen essays for graduate students based on a 1988 symposium. Topics include: animals learning more than we think they do; animals having emotions; and how to improve research methods. Read a publisher review and table of contents at

[www.erlbaum.com/1942.htm]. See an author biography at [www.tulane.edu/~psych/profs/Dachowski.html].

201. Darwin, Charles. *On the Origin of Species by Means of Natural Selection*, reprint ed. New York: Grammercy, 459 pp., 1998 (originally published in 1859). (0517123207) This is one of the most influential books of modern history. Darwin postulated that living creatures evolved in small increments over millenia, based upon genetic improvements that gave survival advantage in a certain ecosystem. It was the strongest statement by science that humans and animals were related by physical properties and by ancestry. Read the whole book online (6th ed.) at [www.literature.org/authors/darwin-charles/the-origin-of-the-species-6th-edition/]. Reviewed at [www1.dragonet.es/users/markbcki/darwin.htm]. See a biography of Darwin at [myhero.com/science/darwin.asp].

202. Darwin, Charles. *The Expression of the Emotions in Man and Animals*, 3rd ed. New York: Oxford University, 448 pp., 1998 (originally published in 1872). (0195112717) This work made Darwin the founder of ethology, the science of studying animal behavior. Since in Darwinian theory man is simply an advanced animal, the importance of the study of animal behavior multiplied into psychology, sociology, anthropology, and other sciences. Despite the fact that his most immediate successors (behaviorists) entirely rejected the premise and evidence listed in this book, Darwin himself believed that animals have emotions like those of humans. Read the whole book online at [www.human-nature.com/darwin/emotion/contents.htm].

203. Dawkins, Marian Stamp. *Through Our Eyes Only? The Search for Animal Consciousness,* reprint ed. New York: Oxford University, 208 pp., 1998. (0198503202) A zoologist from Oxford University searches for biological answers to the question, "Do nonhuman species have consciousness?" The author approaches the subject neutrally, without drawing conclusions. See an article on emotions in chickens at [www.newscientist.com/nsplus/insight/big3/conscious/3a.html].

204. Degrazia, David. *Taking Animals Seriously: Mental Life and Moral Status.* Cambridge: Cambridge University, 302 pp., 1996. (0521567602) "[W]hat sorts of mental capacities we attribute to animals have a great deal to do with how we think they should be treated." (p. 1) "I doubt we should keep any dolphins in aquatic exhibits. For us to provide a comparably good life seems an impossible task in view of their marine habitat and rich social organization." (p. 297) This career philosopher presents a detailed analysis of arguments used in animal rights issues. Read an article by the author at [www.sph.jhu.edu/~altweb/science/meetings/pain/degrazia.htm]. There is

an extensive review of this work at the website [www.hedweb.com/animals/degrazia.htm].

205. Dennett, Daniel Clement. *Kinds of Minds: Toward an Understanding of Consciousness.* New York: Basic, 184 pp., 1996. (0465073514) This philosopher is fun to read: deep in thought, yet simple in words. His final opinion is that thoughts can only come through language, so animals do not have thoughts. He makes interesting distinctions between the human mind and the animal mind in intentionality, sentience, and so on, from an evolutionary perspective. Read articles by the author at [ase.tufts.edu/cogstud/papers/rolelang.htm] and [pp.kpnet.fi/seirioa/cdenn/verblang.htm]. See a related article at [www.newscientist.com/nsplus/insight/big3/conscious/day1a.html].

206. DeWaal, Frans. *Good Natured: The Origins of Right and Wrong in Humans and Other Animals*, 3rd ed. Cambridge: Harvard University, 296 pp., 1997. (0674356616) This expert on primates discusses how morality works among the monkeys. There are surprising evidences of humanlike ethics (conscience, empathy, and sympathy to name a few) among social groups of primates. Article by the author at [www.emory.edu/LIVING_LINKS/a/handclasp.html]. Reviewed at [www.montelis.com/satya/backissues/dec96/bookreview2.html]. This is a fascinating book.

207. Diamond, Jared. *The Third Chimpanzee*. New York: Harper, 416 pp., 1993. (0060984031) This book was also titled *The Rise and Fall of the Third Chimpanzee* in some editions. A UCLA medical professor (winner of a Pulitzer Prize) tells how similar we are to apes. He also says that the choice of humanity to abandon the practice of hunting/gathering in favor of cultivated agriculture has doomed our planet. Read an article by the author at [www.edge.org/3rd_culture/diamond/diamond_p1.html]. See the table of contents at [www.tranby.demon.co.uk/Books/Diamond_RiseFall.htm].

208. Dodman, Nicholas. *The Cat Who Cried for Help: Attitudes, Emotions, and the Psychology of Cats.* New York: Bantam, 256 pp., 1999 (originally published in 1997). (0553378546) The author (a pharmacology professor at Tufts University) has written a similar book about dogs. Why do cats act in such and such ways, and how can human owners of pet cats modify their behavior? Reviewed at [www.bkstore.com/tufts/fac/dodman2.html] and [www.sheba.com/hVV41asd/sheba/reference/book10.html].

209. Dugatkin, Lee Alan. *Cheating Monkeys and Citizen Bees: The Nature of Cooperation in Animals and Humans.* New York: Free Press, 256 pp., 1999. (0684843312) A biologist studies animals in family dynamics, mutual teamwork, universal altruism, and reciprocal transactions. It was

written for a general audience. See the author's resume at [athena.Louis ville.edu/~laduga01/CV.html]. Reviewed at [www.newscientist.com/ns/ 19990213/review.html] and [www.bookideas.com/reviews/science/cheat ingmonkeys.htm].

210.Dugatkin, Lee Alan. *Cooperation Among Animals: An Evolutionary Perspective*. Monographs in Ecology and Evolution series. New York: Oxford University, 240 pp., 1997. (0195086228) Cooperative ventures include the major topics of grooming, foraging, territoriality, and group security. Read an article by the author at [www.sciam.com/1998/0498issue/ 0498dugatkin.html]. Reviewed at [insect-world.com/main/pub/oxford.html #cooperation].

211.Eaton, Randall L. *The Orca Project: A Meeting of Nations, An Anthology*. Enterprise, OR: Sacred Press, 228 pp., 1998. (0966369602) This book is a bit mystical, but wonderful to read and quite interesting. The anecdotal evidence offered about dolphin and whale intelligence and communication is, if true, stunning. It also includes the best short arguments for NOT keeping dolphins and whales in captivity that I have read (chapters 11 and 12).

212.Fichtelius, Karl Eric, and Sjolander, Sverre. *Smarter than Man? Intelligence in Whales, Dolphins, and Humans*. New York: Ballantine, 178 pp., 1974. (0394481496) Using the presupposition that brain size (in proportion to body mass) is key, the authors say that dolphins and whales are at least as intelligent as humans. Of course, in recent years this presupposition about brain size has been largely discredited. See a bibliography on dolphin intelligence at [users.cybercity.dk/~kam2079/ refguide%20dan.htm].

213.Fletcher, David J. C., and Michener, Charles D., eds. *Kin Recognition in Animals*. New York: Wiley, 476 pp., 1987. (0471911992) Evidence is collected to show that the animals' ability to recognize their kin tends to favor the survival of their group, but not the individuals themselves. See a biography of Michener at [ron.nhm.ukans.edu/ksem/people/current/ emeritii/michen.htm]. Read a related article at [rana.uqam.ca/recon2.htm]. Kin recognition is probably not a proof of consciousness or sentience by itself, since it is found even among the lower insects, which are not generally believed to be intelligent in a mammalian sense.

214.Fogle, Bruce. *The Dog's Mind: Understanding Your Dog's Behavior*, 5th ed. New York: Howell, 224 pp., 1992 (originally published in 1990). (0876055137) See a good information site about dogs at [www.geocities. com/Heartland/Prairie/1615/dogs.html].

215. Forrester, Mary Gore. *Persons, Animals, and Fetuses: An Essay on Practical Ethics*. Philosophical Studies series vol. 66. Dordrecht, The Netherlands: Kluwer, 1996. (0792339185) The author is a nurse and philosopher.

216. Fouts, Roger. *Next of Kin: What the Chimpanzees Have Taught Me About Who We Are*. Long Beach, CA: Daniel M. Barber, 407 pp., 1998. (06881-4862x) This scientist writes warmly about his decades of work with primates doing language studies, including the famous chimp Washoe. He says that these apes are abused, and that chimps are nearly extinct in the wild. He has gone from experimenter to animal rights activist. He also says that while primates cannot talk, they do use nonverbal communication effectively. See the author's homepage at [www.animalnews.com/fouts/index.html]. Read an interview with the author at [www.montelis.com/satya/backissues/dec96/keeper.html]. Reviewed at [www.montelis.com/satya/backissues/oct97/one_of_us.html].

217. Fox, Michael W. *Concepts in Ethology: Animal Behavior and Bioethics*, 2nd ed. New York: Krieger, 139 pp., 1998 (originally published in 1974). (1575240440) The author is president of the Humane Society of the U.S. See several other books by Michael W. Fox in other chapters.

218. Frisch, Karl von. *The Dancing Bees: An Account of the Life and Senses of the Honey Bee*. London: Methuen, 183 pp., 1955. This Austrian zoologist, along with Konrad Lorenz and Nikolaas Tinbergen, received the Nobel Prize for Medicine in 1973 for work on behavior patterns in animals. Von Frisch discovered and studied the dance language of the bees, which tells other members of the hive where to locate specific food sources. He is considered to be one of the fathers of modern ethology. See a brief autobiography at [www.nobel.se/laureates/medicine-1973-1-autobio.html]. See a good site about ethologists at [instruct1.cit.cornell.edu/courses/bion420.07/Ethology.html].

219. Gadagkar, Raghavendra. *Survival Strategies: Cooperation & Conflict in Animal Societies*. Cambridge: Harvard University, 192 pp., 1998. (06741-70555) Uses evolutionary thinking to discover why animals risk personal harm to themselves in seeking to protect their social group. Creatures cooperate with each other because their lives are extended (due to less conflict) and the process of finding mates becomes easier. The author works at the Indian Institute of Science, and his page can be seen at [ces.iisc.ernet.in/hpg/ragh/]. See the table of contents at [www.hup.harvard.edu/f97books/f97_catalog/survival_strategies.html].

220. Gallistel, Charles R., ed. *Animal Cognition*. Cambridge: MIT, 211 pp., 1992. (0262570890) A study reprinted from the International Journal of Cognitive Science (37:1-2 in 1990). The author works in the UCLA Psychology Department. The neural basis of learning and motivation. Animal cognition is related to understanding of space, time, and numbers. See a biography of the author at [infosys.psych.ucla.edu/Faculty/Galliste/].

221. Gardner, R. Allen, et al., eds. *Teaching Sign Language to Chimpanzees*. Albany: State University of New York, 324 pp., 1989. (0887069657) Studies the extensive research done in earlier decades with Washoe the chimp; Gardner was one of the original researchers. A good site on this topic is at [math.uwaterloo.ca/~dmswitze/apelang.html]. Read an article by the author at [www.garysturt.free-online.co.uk/gardner.htm].

222. George, Dick. *Ruby, the Painting Pachyderm of the Phoenix Zoo*. New York: Delacorte, 48 pp., 1995. (0385321007) Written for juveniles. Read related articles at [thaifocus.com/elephant/news/story_b.htm] and [fort worthzoo.com/paint.html].

223. Gill, Jerry H. *If a Chimpanzee Could Talk: And Other Reflections on Language Acquisition*. Tucson: University of Arizona, 163 pp., 1997. (081-6516685) Evidence that language acquisition comes more from social interactions than from simple intelligence. Read an excerpt at [www. uapress.arizona.edu/samples/sam1011.htm]. See a related site at [www. pigeon.psy.tufts.edu/psych26/language.htm].

224. Glass, Jay. *The Animal Within Us: Lessons about Life from Our Animal Ancestors*. Corona del Mar, CA: Donington, 156 pp., 1998. (0966053664) The author earned a Ph.D. in neurobiology and psychology, then became a businessman. He analyzes similarities between man and animals, and the origins of emotions. "These insights come from an appreciation of the influence that patterns of behavior we have inherited from our animal ancestors have over our own behavior." (p. ix)

225. Goodall, Jane. *Through a Window: My Thirty Years with the Chimpanzees of Gombe*. Boston: Houghton Mifflin, 268 pp., 1990. (0395599253) This famous biologist has been studying chimpanzees for decades. This work is the sequel to "In the Shadow of Man," following up where she left off with the chimps who live on the shores of Lake Tanganyika, Africa. Read an article by the author at [www.montelis.com/satya/backissues/dec96/ waving.html]. See a biography of Goodall at [www.sjsu.edu/depts/ Museum/goodal.html]. See a site dedicated to saving chimps and gorillas in Africa at [goldray.com/bushmeat/].

226.Gould, James L. *The Animal Mind.* Scientific American Library, volume 51. San Francisco: Freeman, 236 pp., 1994. (0716750465) The author says that the difference between the mental capabilities of man and animal is one of degree and not of kind. Read an excerpt at [www.pigeon.psy.tufts.edu/psych26/kohler.htm]. Reviewed at [members.aol.com/amanitae/ces/books/ani-mind.html].

227.Grandin, Temple, and Deesing, Mark J. *Genetics and the Behavior of Domestic Animals.* San Diego: Academic, 384 pp., 1998. (0122951301) The author is also an expert on autism in humans. Read chapter one at website [www.grandin.com/references/genetics.html]. See an article by the author on animal thinking at [grandin.com/references/thinking.animals.html].

228.Griffin, Donald R. *Animal Minds.* Chicago: University of Chicago, 310 pp., 1992. (0226308634) This pioneer of modern cognitive ethology gives evidence of animal cognition. Animals use signs that are sophisticated and similar to language. This book is a followup to *The Question of Animal Awareness* (published in 1976). The author is a zoology professor at Harvard University. Reviewed at [www.psyeta.org/sa/sa2.1/crist.html]. See a related article at [grimpeur.tamu.edu/~colin/Papers/eor.html].

229.Gucwa, David, and Ehmann, James. *To Whom It May Concern: An Investigation of the Art of Elephants.* New York: Norton, 248 pp., 1985. (0393022404) Regarding the artistic efforts of Siri, a captive Asian elephant, with many illustrations of her paintings and drawings. Read a related article at [animals.co.za/info/causes/entertainment/babyelephants/elephantartmusic.html]

230.Harre, Horace Romano, and Lamb, Roger, eds. *The Dictionary of Ethology and Animal Learning.* Cambridge: MIT, 171 pp., 1986 (originally published in 1983). (0262580764) Harre is a philosophy professor at Oxford University.

231.Hart, Stephen. *Language of Animals.* Scientific American Focus series. New York: Henry Holt, 128 pp., 1996. (080503840x) The Foreword is written by Frans de Waal. The book includes analyses of birds, dogs, insects, and squid. Reviewed at [www.physics.helsinki.fi/whale/intersp/pages/book1.html] and [calvino.biology.emory.edu/signalling/Hart.html].

232.Hauser, Marc D. *The Evolution of Communication.* Cambridge: MIT, 776 pp., 1996. (0262581558) A Harvard psychology professor wrote this scholarly work on the communication of insects, frogs, birds, primates, and humans. Read an interview with the author at [www.edge.org/3rd_culture/

hauser/hauser_index.html]. Reviewed at [ilex.cc.kcl.ac.uk/srb/srb/comm. html].

233. Hearne, Vicki. *Animal Happiness*. New York: Harper Collins, 238 pp., 1994. (0060190167) Hearne has been an animal trainer for more than 25 years. These are stories showing animal joy and human joy in them. Read a brief article by the author at [www.webexchange.net/petadopt/15hear. htm]. See a clever related article at [www.psyeta.org/sa/sa4.2/holbrook. html].

234. Heinrich, Bernd. *Mind of the Raven: Investigations and Adventures with Wolf-Birds*. New York: Cliff Street Books, 380 pp., 1999. (0060174471) The author is a biology professor at the University of Vermont. He adopted baby ravens to observe them closely, and then followed them into their wild habitats. Ravens seem to be remarkably intelligent. "I have become skeptical that the interpretations of all ravens' behavior can be shoehorned into the same programmed and learned responses and categories as those of bees." (p. xix) Read a biography, an interview, and articles by the author at [www.pbs.org/saf/3_ask/archive/bio/93_heinrich_bio.html].

235. Hepper, Peter G., ed. *Kin Recognition*. Cambridge: Cambridge University, 457 pp., 1991. (0521372674) How do animals (from amoebas to man) recognize their relatives? See a biography at [www.psych.qub.ac.uk/ People/Hepper.html]. Read a related article at [www.mat.auckland.ac.nz/ ~king/Preprints/book/socio/kin/kinship.html].

236. Hoage, R. J., and Goldman, Larry, eds. *Animal Intelligence: Insights into the Animal Mind*. National Zoological Park Symposia for the Public series. Washington, DC: Smithsonian, 207 pp., 1986. (0874745411) Essays by several authors (including Sue Rumbaugh, Benjamin Beck and Carolyn Ristau) from a 1983 symposium. "[A]n exciting overview of the current research and theories relating to animal intelligence." (p. 11) See related articles at [www.msu.edu/user/marablek/whal-int.htm] and [www.might mall.com/1st3seconds/shakespr.htm].

237. Honore, Erika K., and Klopfer, Peter H. *A Concise Survey of Animal Behavior*. New York: Academic, 186 pp., 1990. (0123550653) A veterinarian and zoologist (Klopfer, from Duke University) discuss the history of behavior studies. See a great related site at [cas.bellarmine.edu/tietjen/ Ecology/behavioral_ecology_and_evolution.htm].

238. Houck, Lynne D., and Drickamer, Lee C., eds. *Foundations of Animal Behavior: Classic Papers with Commentaries*. Chicago: University of Chicago, 842 pp., 1996. (0226354571) Forty-four facsimiles of classic

works in ethology, written by dozens of experts like Charles Darwin, Robert Hinde, Nikolaas Tinbergen, Karl Von Frisch, Konrad Lorenz, and Donald Griffin. This is an excellent one-volume collection of important animal behavior ideas by the original writers themselves.

239.Howard, Carol J. *Dolphin Chronicles: A Fascinating, Moving Tale of One Woman's Quest to Understand and Communicate with the Sea's Most Mysterious Creatures.* New York: Bantam, 304 pp., 1996. (0553377787) A wonderful account of a two-year dolphin study, written by a graduate student of psychology. It is simple enough for the layman but deep enough for a scholar, in describing behaviors and puzzles about dolphin intelligence. This also provides a good discussion of the issues of keeping dolphins in captivity. Read related articles at [library.advanced.org/17963/index-1.shtml] and links at [www.starrsites.com/acsmb/acslinks.htm].

240.Jacobs, Merle E. *Mr. Darwin Misread Miss Peacock's Mind: A New Look at Mate Selection in Light of Lessons from Nature* (book and video). Nature Books, 272 pp., 1999. (0966591615) A critique of the common view that animals select their mates based on attractiveness. Actually the author (zoology professor at Goshen College) wonders whether animals are even capable of any ideas about beauty. See the book's homepage with links, and Chapter One at [www.goshen.edu/~merleej/book/index.shtml]. Reviewed at [bookflash.com/releases/100117.html].

241.Kennedy, John S. *The New Anthropomorphism.* Cambridge: Cambridge University, 194 pp., 1992. (0521422671) This expert on insect biology offers essays on behavior concepts that have connotations of being anthropomorphic (incorrectly attributing human emotions or qualities to animals). Reviewed at [www.psyeta.org/sa/sa2.1/crist.html]. See a related article at [www.arkanimals.com/WildSide/AniMind.html].

242.King, Barbara J. *The Information Continuum: Evolution of Social Information Transfer in Monkeys, Apes, and Hominids.* Santa Fe, NM: School of American Research (SAR), 166 pp., 1994. (0933452403) How language may have evolved. The author is an anthropology professor at the College of William & Mary. Reviewed at [www.univie.ac.at/Wissenschaftstheorie/srb/srb/face.html].

243.King, Barbara J. *The Origins of Language: What Nonhuman Primates Can Tell Us.* Advanced Seminar series. Santa Fe, NM: School of American Research (SAR), 450 pp., 1999. (0933452608) Different views on language and evolution. The author concludes that language is not unique to humans. See a bibliography of recent primate studies at [pubpages.unh.

edu/~jel/512/primate_bib.html] and read a related article at [crl.ucsd.edu/newsletter/4-4/Article1.html].

244.Klopfer, Peter H. *Politics and People in Ethology: Personal Reflections on the Study of Animal Behavior.* Lewisburg, PA: Bucknell University, 161 pp., 1999. (0838754058) This long-time behavioral ecologist remembers many decades of animal behavior theories and theorists, and discusses the impact of politics on the research done in the field. Scientists always claim to be unbiased and objective, but both their funding and their advancement in the field is dependent upon pleasing the authorities and scientists under whom they work.

245.Koob, George, et al., eds. *Animal Models of Depression.* New York: Birkhauser, 295 pp., 1989. (081763407x) The author is a professor of neuropharmacology at the Scripps Research Institute in California. If animals are useful test subjects in determining causes and cures for depression, does it not follow that they have similar natures to that of humans? See one of his experiments at [www.newswise.com/articles/1998/6/ANXIETY.MSU.html]. Read a related article at [www.trauma-pages.com/yehuda93.htm].

246.Lilly, John Cunningham. *Communication Between Man and Dolphin: The Possibilities of Talking with Other Species.* New York: Crown, 269 pp., 1978. (0517530368) This controversial scientist wrote several books on dolphin intelligence. He reportedly gave dolphins LSD narcotics in some experiments. This work represents more than a decade of research, though much of it has been discounted by peers in the scientific community because of his unorthodox methods and conclusions. See his web page with dolphin links and articles at [www.garage.co.jp/lilly/cetacean.html].

247.Lorenz, Konrad Z. *King Solomon's Ring: New Light on Animals Ways.* New York: Mentor, 215 pp., 1991 (originally published in 1932). (0451628314) The author shared a Nobel Prize in Medicine in 1973 with Nikolaas Tinbergen and Karl von Frisch. Lorenz was a famed bird specialist and early ethologist, whose reputation was somewhat tarnished by his relationship with the Nazi party. This book is easy to read and full of insights, showing what ethology is about. Read an autobiography at [www.nobel.se/laureates/medicine-1973-2-autobio.html]. See an article about Lorenz' theories at [141.163.90.14/year1/lorenzexplanation.html].

248.Lutts, Ralph H. *The Nature Fakers: Wildlife, Science and Sentiment.* Golden, CO: Fulcrum, 255 pp., 1990. (1555910548) The author teaches environmental studies at Hampshire College. An interesting study regarding John Burroughs, who accused many researchers and writers of romanticizing and exaggerating stories about animal abilities, in the year

1903. Outlandish accounts end up hurting the animal rights movement since they give a "black eye" to research work. Read a publisher review and excerpt at [fulcrum-books.com/html/nature_fakers.html].

249. Mack, Arien, ed. *Humans & Other Animals*. Columbus: Ohio State University, 450 pp., 1999. (0814250173) Edited essays from a 1995 conference on Human/Animal Interactions by the New School for Social Research. It includes chapters by S. J. Gould, Daniel Dennett, Andrew Rowan, Sue Rumbaugh, Vicki Hearne, and Juliet Clutton-Brock. Read an article by the author at [psyche.cs.monash.edu/v5/psyche-5-03-mack.html]. See a publisher review and the table of contents at [www.ohiostatepress.org/S99/MACHUM.htm].

250. Masson, Jeffrey Moussaieff. *Dogs Never Lie About Love: Reflections on the Emotional World of Dogs*. New York: Random House, 274 pp., 1998. (0609802011) The author has been unjustly attacked as being a "mere" Sanskrit language scholar, as if only biologists can write about animals. However, criticism that he overstated his evidences may be valid: the author apparently has never owned any pets, and only bought a few to write this book, reportedly. Of course, anyone can make "reflections." Read an article by the author at [www.earth.org.hk/dogweep.html]. Reviewed at [asa.calvin.edu/ASA/book-reviews/6-96.htm].

251. Masson, Jeffrey Moussaieff. *The Emporer's Embrace: Reflections on Animal Families and Fatherhood*. New York: Pocket Books, 304 pp., 1999. (0671020838) An interesting comparative study on animal families as compared to human families. Includes discussions of the advantages of monogamy and adoption.

252. Masson, Jeffrey Moussaieff and McCarthy, Susan. *When Elephants Weep: The Emotional Lives of Animals*. New York: Delacorte, 291 pp., 1995. (0385314256) A very popular book of anecdotal stories about animal capacities for love, joy, anger, fear, shame, compassion, loneliness, and other emotions. Readers either loved it or hated it in reviews. Read an interview with McCarthy at [arrs.envirolink.org/ar-voices/mccarthy.html]. Reviewed at [www.psyeta.org/ sa/sa4.1/moore.html]. See an article on animal grief at [www.abilnet.com/norclub/INTC-NTCI/HowPetsGrieve01.html].

253. McCrone, John. *The Ape that Spoke: Language and the Evolution of the Human Mind*. New York: Avon, 288 pp., 1992. (068810326x) Non-technical accounts of theories about why intelligence and communication are such strong survival mechanisms in evolution. Read an article by the

author at [cogweb.english.ucsb.edu/CogSci/DynamicBrain.html]. See a related article at [dspace.dial.pipex.com/jcollie/els/index.htm].

254. McGrew, William C., et al. *Great Ape Societies.* Cambridge: Cambridge University, 328 pp., 1996. (0521554942) 21 essays on gorillas, chimps, orangutans, and bonobos. Of particular interest are the commonalities shared with humans: high intelligence, social groupings, lengthy adolescent periods, and omnivorous diets. Read a review of an earlier work by the author at [www.univie.ac.at/Wissenschaftstheorie/srb/srb/roots.html]. See a list of study sites (created by McGrew) at [weber.ucsd.edu/~jmoore/apesites/ApeSite/html].

255. McIntyre, Joan, ed. *Minds in the Water: A Book to Celebrate the Consciousness of Whales and Dolphins.* New York: Scribner's, 240 pp., 1974. (0684139952) 28 essays by Farley Mowat and others, giving evidence of consciousness among cetaceans and pleading for their preservation. See an article on self-awareness in dolphins at [planet-hawaii.com/earthtrust/delbook.htm].

256. Milani, Myrna M. *Body Language and Emotions in Dogs.* New York: Quill, 283 pp., 1993 (originally published in 1986). (0688128416) The author also wrote a similar book on cats. One interesting assertion is that human pet owners must be careful, as they often misinterpret dog behavior as being aggressive or friendly when it may be neither, and the owners may be reinforcing negative behaviors rather than good ones. Read an article about body language in rabbits at [www.whiskerwisdom.com/communic.htm].

257. Milne, Lorus Johnson, and Milne, Margery. *The Behavior and Learning of Animal Babies.* Chester: Globe Pequot, 162 pp., 1988. (0871066149) Read an article about animal learning at [www.biology.ucsc.edu/~barrylab/classes/animalbehavior/LEARNING.HTM].

258. Mitchell, Robert W., et al., eds. *Anthropomorphism, Anecdotes and Animals.* Philosophy and Biology series. Albany: State University of New York, 518 pp., 1997. (0791431266) Debates on the usefulness and dangers of anthropomorphism (attributing human characteristics to animals). Very detailed studies (on both sides of the argument) of human attribution of intelligence and emotions to animals. See a brief review at [www.sunypress.edu/sunyp/backads/html/mitchellthompsonmilesanthro.html].

259. Mitchell, Robert W., and Thompson, Nicholas S., eds. *Deception: Perspectives on Human and Nonhuman Deceit.* SUNY Series in Animal Behavior. Albany: State University of New York, 388 pp., 1986.

(0887061079) This psychologist says that there are four levels of deceit: two are involuntary (instinctual), and two are voluntary (by choice). Some animals only practice the involuntary deceptions, but other animals choose or plan their deceptions. See the author's ideas on multiple links beginning at [www.a3.com/myself/truthppr.htm #TofC10]. Read related articles at [www.newscientist.co.uk/ns/980214/features.html] and [....features2.html]. Is animal deception a moral question? Is this "sin?" Does deception does imply some sort of advanced intelligence?

260.Morris, Desmond. *Animal Watching: A Field Guide to Animal Behaviour*. New York: Warner, 169 pp., 1991. (0517083388) See a related article at [asci.uvm.edu/bramley/Behavior.html].

261.Mortenson, F. Joseph. *Whale Songs & Wasp Maps: The Mystery of Animal Thinking*. New York: Dutton, 178 pp., 1987. (0525244425) "[T]he drab, dark mechanistic world of the behaviorist is beginning to be illuminated by the bright light of animal awareness." (dust jacket flap) This is a marvelous work about consciousness and intelligence in species. The author has a Ph.D. in experimental psychology. Read a related article at [www-und.ida.liu.se/ ~danji745/HTML-version.html].

262.Morton, Eugene S., and Page, Jake. *Animal Talk: Science and the Voices of Nature*. New York: Random House, 273 pp., 1992. (039458337x) A study which focuses largely on bird communication. The authors conclude that human communication differs from that of animals only in degree, not in kind. See a biography of Morton at [www.inform.umd.edu/EdRes/ Colleges/LFSC/ life_sciences/.WWW.zoology/morton.htm]. Read a related article at [www.ualr.edu/~lmwilliams/cole.html].

263.National Audobon Society. *If Dolphins Could Talk* (video). National Audobon Society Specials series. Van Nuys, CA: Vestron Video, 1991. (1556589387) 60-minute video, hosted by Michael Douglas. There are many good videos about dolphins, which you can also see frequently on the cable channel "Animal Planet." Dolphins are one of the animals frequently studied by scientists because of their complex communication abilities. Does this constitute language? See a related article at [www.loop.com/ ~dacs/dolphin.htm].

264.Nollman, Jim. *The Charged Border: Where Whales and Humans Meet*. New York: Henry Holt, 249 pp., 1999. (0805055231) The author has been attempting to communicate with whales and dolphins by using musical signals underwater. This book focuses on a variety of attitudes that people have toward whales, written for a popular audience (not scholars). Read an excerpt at [www.physics.helsinki.fi/whale/intersp/pages/behalf.html].

265. Noske, Barbara. *Humans and Other Animals: Beyond the Boundaries of Anthropology*. London: Pluto Press, 244 pp., 1989. (1853050547) Reviewed at [www.psyeta.org/sa/sa2.1/crist.html]. Read an article by the author at [www.psyeta.org/sa/sa1.2/noske.html].

266. Novak, Melinda A., and Petto, Andrew J., eds. *Through the Looking Glass: Issues of Psychological Well-Being in Captive Non-Human Primates*. Washington, DC: American Psychological Association, 298 pp., 1991. (155798087x) 23 articles taken from a 1988 conference held in Boston. See the table of contents at [www.apa.org/books/glasst.html]. Find a bibliography at [www.animalwelfare.com/Lab_animals/biblio/]. Animals have psychological needs as well as physical ones. Many animals in zoos, like elephants, receiving training to keep them from becoming bored!

267. Owings, Donald H., and Morton, Eugene S. *Animal Vocal Communication: A New Approach*. Cambridge: Cambridge Univ., 284 pp., 1998. (0521324688) A scholarly work that says that animals often use vocal communication to "manage" or manipulate others to act to help them, though "assessors" may ignore these attempts at management. Read a publisher review at [www.cup.cam.ac.uk/scripts/webbook.asp?isbn=05213 246688].

268. Parker, Sue Taylor, et al., eds. *Self-Awareness in Animals and Humans: Developmental Perspectives*. Cambridge: Cambridge University, 464 pp., 1994. (0521441080) Much of this study centers around primates and the debate over "mirror self-recognition." Do monkeys really know that they see themselves in the mirror, or are they simply reacting to "that other monkey" they see in the mirror? Read an excerpt [planet-hawaii.com/ earthtrust/delbook.html]. See a related article at [www.newscientist.com/ ns/971025/ lastword.html].

269. Parker, Sue Taylor, and Gibson, Kathleen Rita, eds. *"Language" and Intelligence in Monkeys and Apes: Comparative Developmental Perspectives*. Cambridge: Cambridge University, 590 pp., 1994 (originally published in 1991). (0521459699) Articles from authors in several countries, including Japan, Spain, Italy, France, Canada, and the United States, on intelligence in various animals including primates and parrots. See the table of contents at [www.cup.cam.ac.uk/scripts/webbook.asp?isbn=05214 59699]. See a related site with articles and links at [www.cwu.edu/ ~cwuchci/main.html].

270. Parker, Sue Taylor and McKinney, Michael L. *Origins of Intelligence: The Evolution of Cognitive Development in Monkeys, Apes, and Humans*. Baltimore, MD: Johns Hopkins University, 404 pp., 1999. (0801860121)

How Darwinian evolution produced successful adaptations in the primates and humankind. Taylor is a comparative psychologist; McKinney is an evolutionary theorist. The book is quite deep in theory, but readable for students.

271. Pearce, John M. *Animal Learning and Cognition,* 2nd revised ed. Hove, England: Psychology Press, 352 pp., 1997. (0863774334) The author is the editor of the *Quarterly Journal of Experimental Psychology*. See a publisher review at [www.tandfdc.com/PSYPRESS/BKFILES/0863774342.htm]. Read a related article [www.williamcalvin.com/1990s/1997SciEds.htm].

272. Peterson, Dale, and Goodall, Jane. *Visions of Caliban: On Chimpanzees and People*. Boston: Houghton Mifflin, 367 pp., 1994. (0395701007) The authors show the danger of extinction (in the wild) for the chimpanzees in Africa, largely due to deforestation from logging and fuel-fire needs. They ask themselves if chimps can only now be saved in the world's zoos. See Jane Goodall's Institute Center for Primate Studies at [www.cbs.umn.edu/chimp/index.html]. Read an article about Goodall's research at [www.arts.mcgill.ca/programs/anthro/asa/digest/anna.html].

273. Pinecrest, Richard Franklin. *Animal Social Behavior: Index of New Information with Authors, Subjects & Bibliography*. Washington, DC: ABBE, 1993. See a related bibliography at [www.cisab.indiana.edu/ABS/Education/books.html].

274. Premack, David. *Gavagai! Or the Future History of the Animal Language Controversy*. Cambridge: MIT, 155 pp., 1986. (0262160994) The author is a professor emeritus at the University of Pennsylvania. He discusses the arguments of biologists, linguists, and philosophers about the "talking" chimps. Read related articles at [www.ldc.upenn.edu/ myl/lx1/nonhuman.html] and [www.geocities.com/RainForest/Vines/4451/TalkWithChimps.html].

275. Pryor, Karen, and Norris, Kenneth, eds. *Dolphin Societies: Discoveries and Puzzles*. Berkeley: University of California, 397 pp., 1998 (originally published in 1991). (0520216583) 24 articles written by experts and graduate students, about studies of wild and captive dolphins. See the author's website at [www.clickertraining.com/]. Read an article on societal evolution [www.biology.ucsc.edu/~barrylab/classes/animal_behavior/SOCIAL.HTM].

276. Quiatt, Duane, and Reynolds, Vernon. *Primate Behavior: Information, Social Knowledge, and the Evolution of Culture*. Cambridge Studies in

Biological Anthropology Volume 12. Cambridge: Cambridge University, 322 pp., 1993. (0521498325) Anthropologists (Quiatt, University of Denver) look at the social lives in primates and among humans. Reynolds teaches anthropology at Oxford University. Read a related article at [www.cogsci.soton.ac.uk/bbs/Archive/bbs.dunbar.html].

277. Rachels, James. *Created from Animals: The Moral Implications of Darwinism*. New York: Oxford University, 256 pp., 1991. (0192861298) See a publisher review at [www.oup-usa.org/docs/0192861298.html] and a brief biography at [www.uab.edu/philosophy/faculty/rachels/index.htm].

278. Radner, Daisie, and Radner, Michael. *Animal Consciousness,* reprint ed. Buffalo, NY: Prometheus, 253 pp., 1989. (1573921149) These philosophy professors claim that Rene Descartes is not the "beast machine" villain that many animal rights activists think. See a related article at [www. newschool.edu/gf/psy/faculty/humphrey/consciousness.htm].

279. Richards, Robert J. *Darwin and the Emergence of Evolutionary Theories of Mind and Behavior*, reprint ed. Science and Its Conceptual Foundations series. Chicago: University of Chicago, 700 pp., 1989 (originally published in 1987). (0226712001) A study of Darwin and his contemporaries' ideas about behavior and mind in animals, with allusions also to sociobiology and ethology. Read an article by the author at [www.cogsci.princeton.edu/ ~ghh/COG/Richards.txt]. See a related article [www.human-nature.com/ dm/chap3.html] and bibliography [www.dropbears.com/b/broughsbooks/ darwin.htm].

280. Ristau, Carolyn A., ed. *Cognitive Ethology: The Minds of Other Animals*. Hillsdale, NJ: Lawrence Erlbaum, 332 pp., 1991. (0805802517) The presence of language and deception shows awareness and cognition in animals. See an introduction to ethology at [cas.bellarmine.edu/tietjen/ Ethology/ethology_ main.htm].

281. Roberts, William A. *Principles of Animal Cognition*. Boston: McGraw-Hill, 464 pp., 1997. (0070531382) This researcher from the University of Western Ontario has been studying memory in pigeons, rats, and monkeys for decades. He says that animals have both short-term and long-term memory, and they can perform perceptual, relational, and associative concepts. Includes a substantial bibliography. See the author's page at [yoda.sscl.uwo.ca/psychology/faculty/roberts.html].

282. Rogers, Lesley J. *The Development of Brain and Behaviour in the Chicken*. Wallingford, England: CAB International, 288 pp., 1995. (0851989241) A scholarly and detailed work studying chickens in comparison with other

birds. This work has direct significance to the animal rights movement since the chicken (in factory farms) is probably the most frequently eaten animal in the world. See the table of contents at [www.cabi.org/CATALOG/BOOKS/book_detail.asp?ISBN=0851989241]. Read a related article at [www.upc-online.org/ genetic.html].

283.Rogers, Lesley J. *Minds of Their Own: Thinking and Awareness in Animals*. Boulder, CO: Westview, 212 pp., 1997. (0813390656) This doctor of philosophy in ethology questions traditional views on animal consciousness and behavior. Do animals think abstractly? When did consciousness evolve in the evolutionary history of organisms? She concludes that animal intelligence is greater than is commonly believed. An interesting point, also, is that modern science and industry have been breeding animals to be more docile (and therefore perhaps less intelligent) so that they are better in factory farms and laboratories. Read an article about how people view the minds of dogs at [www.psyeta.org/sa/sa3.2/rasmuss.html].

284.Rollin, Bernard E. *The Unheeded Cry: Animal Consciousness, Animal Pain, and Science*, 2nd expanded ed. Ames: Iowa State University, 344 pp., 1998 (originally published in 1989). (081382575x) The author argues for moral rights as the ideal, but since this ideal is not currently attainable, we must use a utilitarian approach. "My hope is that this book will help scientists break the ideological bonds which keep them from ascribing subjective mental states to animals." (p. xii) Reviewed at [www.nsplus.com/nsplus/insight/animal/double.html]. Rollin says that we should take moderate positions, for now, since the ideals cannot be had. Gary Francione and others disagree, saying that we must have all or nothing. Complete polarization without any room for a middle ground is a difficult place to bargain from; both sides tend to dig in their heels.

285.Romanes, George J. *Animal Intelligence*. International Scientific series volume 41. Oxford: Kegan Paul, 520 pp., 1898 (originally published in 1882). An influential early work in comparative psychology and the evolution of mental powers. Read the whole book online at [post.queensu.ca/~forsdyke/romanes1.htm] or excerpts at [www.pigeon.psy.tufts.edu/psych26/romanes.htm].

286.Russon, Anne E., Bard, Kim A., and Parker, Sue Taylor, eds. *Reaching into the Minds of the Great Apes*. Cambridge: Cambridge University, 464 pp., 1996. (0521471680) 25 essays by contributors, including Frans DeWaal and Sarah Boysen. Russon worked with Orangutans in Borneo, and writes that apes can indeed think at symbolic levels. How are ape minds different than human minds? Reviewed at [www.univie.ac.at/

Wissenschaftstheorie/srb/srb/mindful.html]. See also website [www.pbs.org/edens/borneo/orangs.html].

287. Sagan, Carl. *The Dragons of Eden: Speculations on the Evolution of Human Intelligence*, reprint ed. New York: Ballantine, 240 pp., 1989 (originally published in 1977). (0345346297) "The main conclusion arrived at in this work, namely, that man is descended from some lowly-organized form, will, I regret to think, be highly distasteful to many persons." This popular scientist condenses evidence on how animals reason, and on the evolution of intelligence in organisms. This book won a Pulitzer Prize. See a Sagan page with links and articles at [www.blueprint.com.tr/sagan/index.html]. Read a related article at [members.aol.com/jligda/projects/towerch5.htm].

288. Savage, Candace C. *Bird Brains: The Intelligence of Crows, Ravens, Magpies, and Jays.* San Francisco: Sierra Club, 144 pp., 1997. (08715-69566) Lots of photographs and some text documenting the intelligence of 16 species of birds. The overall view is that these birds are highly intelligent. Briefly reviewed at [www.azstarnet.com/~serres/book.html]. See related articles at [www.appi.org/pnews/crow.html] and [www.cages.org/research/pepperberg/index.html]. See also the book by Bernd Heinrich in this chapter.

289. Savage-Rumbaugh, E. Sue, et al. *Apes, Language, and the Human Mind.* New York: Oxford University, 288 pp., 1998. (0195109864) An analysis of Kanzi, a Bonobo chimp who is said to have the language ability of a 2.5-year-old human child. This work generated some controversy for its strong criticisms of opponents. Read an interview with the author at [www.geocities.com/RainForest/Vines/4451/SheTalks.html] and an interview with one of her chimps at [pubpages.unh.edu/~jel/512/chimps/SSR.html]. See her workplace at the Language Research Center (Georgia State University) [www.gsu.edu/~wwwlrc/].

290. Savage-Rumbaugh, E. Sue, and Lewin, Roger. *Kanzi: The Ape at the Brink of the Human Mind.* New York: Wiley, 299 pp., 1994. (047115959x) Detailed study of a Bonobo chimp who is learning human language and communication. Includes a good history of primate language studies. Reviewed at [arrs.envirolink.org/ar-voices/nearly.html] and [pubpages.unh.edu/~jel/SGMonKanzi.html]. See a related article on Rumbaugh at [www.geocities.com/RainForest/Vines/4451/ApesCommunicate.html].

291. Schmajuk, Nestor A. *Animal Learning and Cognition: A Neural Network Approach.* Cambridge: Cambridge University, 340 pp., 1997. (0521450-861) A Duke University professor of psychology and computer science

presents a very scholarly view of animal intelligence. How animals learn, integrating emotive, physiological, and cognitive aspects of their minds. Includes a 3.5-inch floppy disk with a DOS program. See the table of contents at [psych.duke.edu/cnlab/sch21.htm].

292. Scholtmeijer, Marian. *Animal Victims in Modern Fiction: From Sanctity to Sacrifice*. Toronto: University of Toronto, 330 pp., 1993. (0802028322) The author teaches English at Northwest Community College. How authors of fiction (like Jack London, Ernest Hemingway, John Steinbeck, D.H. Lawrence, and Stephen King) describe the victimization and natures of animals. We have very contradictory and confused ideas about animals, shown in writers' attempts to portray them. Read an interesting related article at [www.psyeta.org/sa/sa4.1/johnson.html].

293. Schusterman, Ronald J., et al., eds. *Dolphin Cognition and Behavior*. Comparative Cognition and Neuroscience series. Hillsdale, NJ: Lawrence Erlbaum, 393 pp., 1986. (0898595940) See a biography and bibliography at [natsci.ucsc.edu/acad/oceansci/adjunct/rjsCV.html]. Related articles and bibliographies at [www.scils.rutgers.edu/~roccos/dolphint.html] and [www.rtis.com/nat/user/elsberry/marspec/dolphin.html].

294. Sebeok, Thomas A., and Sebeok, Robert. *The Clever Hans Phenomenon: Communicating with Horses, Whales, Apes, & People*. New York: Annual New York Academy of Sciences, 310 pp., 1983. (0897661141) The title refers to the famous horse Clever Hans, who supposedly could count using mathematics by stomping his hoof. It was soon determined, however, that Hans was getting subtle clues from his master and audience as to when he should quit stomping. This discredited many other animal studies, as scientists could thus discount all anecdotal evidences as being "tainted" by human interference in the animal's behavior.

295. Shapiro, Kenneth Joel. *Animal Models of Human Psychology: Critique of Science, Ethics, and Policy*. Seattle: Hogrefe & Huber, 344 pp., 1998. (088937189x) Scholarly reflections on animal issues. Reviewed at [www.montelis.com/satya/backissues/sep97/mental_torture.html].

296. Shepard, Paul. *The Others: How Animals Made Us Human*. Washington, DC: Island, 374 pp., 1996. (1559634332) Written by a well-known ecologist from Yale University. He says that animals led to the greater intelligence of humankind, but he blasts animal rights ideas, and he loved hunting with a passion. See his biography and obituary at [www.csuhayward.edu/ALSS/ECO/0197/shepard.htm]. Reviewed at [www.montelis.com/satya/backissues/ sep97.shepards_lore.html].

297. Sherman, Paul W., and Alcock, John, eds. *Exploring Animal Behavior: Readings from American Scientist*, 2nd ed. Sunderland, MA: Sinauer, 300 pp., 1998. (0878937625) Thirty essays over a twenty-year period, intended to be a supplement for the book *Animal Behavior* by Alcock. Includes classroom assignments and topics for discussion. Writers include Stephen Jay Gould, Donald Griffin, and E.O. Wilson. See the table of contents at [www.webcom.com/~sinauer/saw/ Titles/Text/sherman.html].

298. Shettleworth, Sara J. *Cognition, Evolution, and Behavior*. Oxford: Oxford University, 704 pp., 1998. (019511048x) The author attempts to combine research from the fields of ecology, ethology, and psychology to arrive at determinations of the mental capacities of animals.

299. Shipman, Wanda. *Animal Architects: How Animals Weave, Tunnel, and Build Their Remarkable Homes*. Mechanicsburg, PA: Stackpole Books, 160 pp., 1994. (0811724042) "Animals are much more than the instinct-driven automatons that science for centuries held them to be." (p. ix) Shipman points out the complexity and variety of homes, built with adaptations to changing environments. Included are prairie dogs, spiders, bears, bees, and stickleback fish. She is an editor from *Wilderness* magazine. Read a related article at [home.worldcom.ch/~negenter/ 081NestbApes_E.html].

300. Short, Charles E., and Van Poznak, Allan, eds. *Animal Pain*. Edinburgh: Churchill Livingstone, 616 pp., 1992. (0878937625) See a brief biography of Short at [www.avma.org/pubinfo/piaward.htm#DietsFido]. Read a related article at [www.rabbit.org/journal/3-10/ pain.html].

301. Skutch, Alexander F. *The Minds of Birds*, reprint ed. College Station: Texas A&M, 184 pp., 1999 (originally published in 1996). (0890967598) The author is a famed ornithologist with a Ph.D. in botany. He says that birds are comparable to non-human mammals in intelligence. They have well-developed memory, including emotions, play behavior, tool use, sense of aesthetics, and use of deceit. He studied birds for more than 60 years. See an article about the author at [www.post-gazette.com/columnists/1998 0827walk3.asp]. Reviewed at [www.calacademy.org/pacdis/issues/spring 97/review.htm].

302. Slater, P. J. B. *Essentials of Animal Behaviour*. Cambridge: Cambridge University, 233 pp., 1999. (0521629969) Designed for undergraduates, the author presents an illustrated study of ethological studies, with particular emphasis on genetic determination of behavioral traits.

303. Smuts, Barbara Boardman, ed. *Primate Societies*. Chicago: University of Chicago, 578 pp., 1986. (0226767167) Read an article by the editor at [www.mc.maricopa.edu/anthro/origins/apeswrath.html]. The book includes a 52-page bibliography. The book is composed of essays by 46 contributors including works by Dorothy Cheney and Robert Seyfarth.

304. Sorabji, Richard. *Animal Minds and Human Morals: The Origins of the Western Debate*. Cornell Studies in Classical Philology, the Townsend Lectures #54. Ithaca, NY: Cornell University, 267 pp., 1993. (08014-82984) The author is a professor of ancient philosophy: he discusses the animal rights debate from Aristotle to modern times. Reviewed at [www.psyeta.org/sa/sa3.2/serpell.html] and [www.ivu.org/books/reviews/animal-minds.html]. See a brief biography at [www.kcl.ac.uk/kis/schools/hums/philosophy/staff/ richards.html].

305. Strum, Shirley C., and Schaller, George B. *Almost Human: A Journey into the World of Baboons*. New York: Random House, 294 pp., 1987. (0393307085) Read an article by Strum at [www.nwf.org/nwf/intlwild/1998/baboon.html].

306. Temerlin, Maurice K. *Lucy: Growing Up Human: A Chimpanzee Daughter in a Human Family*. Palo Alto, CA: Science and Behavior Books, 216 pp., 1975. (0831400455) This was a controversial experiment known as "cross fostering," in which an animal (in this case a chimpanzee) was led to believe that it was human by being raised as a human in a human family. Of course, it is not all that different from the experience of a pet.

307. Terrace, Herbert S. *Nim: A Chimpanzee Who Learned Sign Language*, reprint ed. New York: Columbia University, 303 pp., 1987 (originally published in 1979). (0231063415) Can animals learn language as people do? The author concluded that they lack language, but they do think. He later repudiated his own work. See an article about Nim the chimp at [blackbeautyranch.org/originals/nimffa.html]. Brief related articles can be found at [ego.psych.mcgill.ca/faculty/petitto/ape.html] and [www.ex.ac.uk/~bosthaus/Lecture/nim.htm].

308. Thomas, Elizabeth Marshall. *The Hidden Life of Dogs*. Boston: Houghton Mifflin, 148 pp., 1993. (0395669588) "This is a book about dog consciousness." Reviewed at [www.psyeta.org/sa/sa4.1/mitchell.html].

309. Thomas, Warren D., and Kaufman, Daniel. *Dolphin Conferences, Elephant Midwives, and Other Astonishing Facts about Animals*. Los Angeles: Tarcher, 179 pp., 1990. (0874775876) A collection of interesting physiological and behavioral trivia about animals. Gives readers a taste of the

wonders and variety among animals, which we can hardly understand. See a great life sciences site at [nua-tech.com/paddy]. The Birmingham Zoo has links to many popular animal species ("Animal Omnibus") at [www.birminghamzoo.com/ao/].

310. Tinbergen, Nikolaas. *The Study of Instinct.* New York: Oxford University, 228 pp., 1989 (originally published in 1951). (0198577400) One of the founders of modern ethology, Tinbergen shared the Nobel Prize in Medicine in 1973 with Karl Von Frisch and Konrad Lorenz. Read an autobiography at [nobel.sdsc.edu/laureates/medicine-1973-3-autobio.html].

311. Tobias, Michael, and Mattelon, Kate, eds. *Kinship with the Animals.* Hillsboro: Beyond Words, 337 pp., 1998. (1885223889) "[W]e have expressly sought out extraordinary stories, anecdotes, and the rich, subversive science of encounters between humans and other life forms that might truly expand and accelerate the vocabulary of interspecies dialogue and empathy." (p. xi) The second essay, by Anthony Rose, is simply outstanding.

312. Trefil, James S. *Are We Unique? A Scientist Explores the Unparalleled Intelligence of the Human Mind.* New York: Wiley, 256 pp., 1997. (0471155365) A very readable and enjoyable book. He cites evidence to prove that humans are far more intelligent than animals. See a brief biography at [www.kalmbach.com/Astro/Staff/advisory/trefil.html]. See a related site at [www.fi.edu/qa97/spotlight5/].

313. Vauclair, Jacques. *Animal Cognition: An Introduction to Modern Comparative Psychology.* Cambridge: Harvard University, 256 pp., 1996. (0674037030) The author is director of research at a cognitive studies laboratory in France. He analyzes the psychology of primates, dogs, mice, dolphins, and bees. See a publisher review at [hupress.harvard.edu/S96_Books/S96_Long/animal_cog.html]. Read a related article at [lucs.fil.lu.se/staff/Christian.Balkenius/Thesis/Chapter02.html].

314. Walker, Stephen F. *Animal Learning.* Introduction to Modern Psychology series. New York: Routledge, 437 pp., 1987. (0710204825) The author is a psychology professor at the University of London. He rejects strict behaviorist theories of animal thinking. "[I]t makes sense to suppose that awareness and mental organisation occur in animals, without the involvement of language." See a brief biography at [www.psyc.bbk.ac.uk/staff/sfw.html]. Read a related article at [www.nlu.edu/~chutto/twoohone/201C09.html].

315. Wallman, Joel. *Aping Language*. Themes in the Social Sciences series. Cambridge: Cambridge University, 1992 (originally published in 1984). (0521404878) The author critiques animal language experiments, saying that primates are too different from humans to even be tested. Wallman says that language is uniquely a human trait. Read a related article at [www.psych.auckland.ac.nz/Psych/teaching/461.250/TEACH/ProblemSolving/Problem.html].

316. Wasmann, Eric. *Comparative Studies in the Psychology of Ants and of Higher Animals*. St. Louis, MO: B. Herder, 200 pp., 1995 (originally published in 1918). This is a classic in animal behavior. Ants and other insects do have fairly complex behavior patterns, which make distinguishing instinct from intelligence more difficult.

317. Weiskrantz, Lawrence, ed. *Thought Without Language*. Oxford: Clarendon, 533 pp., 1988. (0198521774) Read an article about the author's views at [www.unc.edu/~eckerman/P22day26.html]. See a related article at [www.ditext.com/chrucky/chru-5.html].

318. Wemelsfelder, Francoise. *Animal Boredom: Toward an Empirical Approach of Animal Subjectivity*. Utrecht, the Netherlands: Elinkwijk, 195 pp., 1993. Read a condensed version at the website [www.psyeta.org/hia/vol8/wemelsfelder.html]. Animals in zoos are often trained, not so much to generate fun exhibits for human-onlookers, but because intelligent creatures (like elephants) become bored and often depressed when there are no challenges in their environments.

319. Wenner, Adrian M. *Anatomy of a Controversy: The Question of a Language Among Bees*. New York: Columbia University, 399 pp., 1990 (originally published in 1971). Read a related article at [www.geocities.com/Athens/Delphi/8309/langfac.html]. See a webpage with useful information and links about bee language at [ag.arizona.edu/pubs/insects/ahb/lsn14.html].

Chapter 3

Fatal Uses of Animals

Human beings kill an incomprehensible number of animals each day for a number of reasons. Most creatures are slain for what people would call "good cause," that is, food.

In general, Americans are unfamiliar with other cultures, and they are surprised to learn that the peoples of other nations consume much less meat. The modern first-world tradition of eating meat three meals per day only became possible with the advent of the Industrial Revolution in the nineteenth century.

Certainly, humankind has been killing and consuming animals for a very long time. Evolutionary theory says that human societies moved from being hunter-gatherers to cultivators of agriculture only a few thousand years ago. Primates are not simple vegetarians; they are omnivores, eating both plant life and animal life, like humans.

Physiologically, humans are not well designed for a solely carnivorous diet. Our teeth and digestive systems could not cope with such a diet. The simplest debate on meat-eating arises amid the native or indigenous peoples, like the Inuit of the Arctic Circle, who cannot survive without meat. The land is not arable for farming, and transport is scarce and expensive. For thousands of years, native peoples have taken pride in their ability to fend for themselves — a very independent lifestyle. Most of us would not fault a man or woman from killing and eating an animal if the alternative was starvation. On the other hand, these people do choose to live in a barren place, where animal killing is the only way to survive.

The next question is, should we begrudge people for killing animals solely for economic reasons? Many natives fund their isolated lifestyles by selling furs and pelts, oil boiled from blubber, or tusks for ivory trinkets. Is wildlife to be used as a financial resource?

Factory farming is the key change in the twentieth century that has led to the explosion in meat-quantity consumption by the societies living in first-world nations. It would be impossible to raise enough meat for the current demand using traditional family farm methods, due to lack of land and labor. So in modern animal agriculture, animals are raised in confined and crowded spaces, with little or no freedom of movement, and they are injected with hormones and antibiotics to ensure quicker fattening and fewer diseases. The increase in meat production, of course, leads to an increase in the number of animals that slaughterhouses must process. Critics claim that factory farming causes major environmental and medical problems, not to mention increased stress and disease among the animals. The so-called Mad Cow Disease, which killed dozens in Great Britain, was related to this factory farming method of production, since the animal waste-products not to be consumed by humans are processed into animal food.

Another realm in which modern industrial methods of food production have had a major impact is the world's oceans. Whaling, the killing of whales for oil and other products, had a negligible impact on whale populations until the nineteenth century, when mechanized ships and equipment became standard. Explosive harpoons and steamships gave whalers the power and resources to take larger numbers of whales. Furthermore, new practices, like drift-netting (which can stretch for miles) and long-lining have greatly reduced the populations of "food fish." A related practice called finning occurs when fisherman simply cut off the fins of a shark or other fish, because only the fins are valuable. The still-living creature is tossed back into the sea, to be consumed by other sharks.

Clothing for humans often requires the death of an animal. Leather and furs come from animal hides. In the past, furs were primarily gleaned from wild animal populations, but trapping completely decimated the wild populations of fur-bearing animals. Now most furs come from farm-raised animals. Trapping is controversial for a number of reasons. For one thing, traps cannot discriminate between fur-bearers and non-fur-bearers, so traps often snare the wrong animals. Also, if the trapper does not frequently check the traps, a snared animal may be held captive for a long time without food or water or shelter. Fur has become less popular in modern times, as there has been a strong anti-fur campaign run by animal rights activists around the world. One practice that has largely fallen out of favor is the killing of birds for their exotic feathers to put on ladies' hats. But a similar problem, that of amateur ornithologists who collect bird egg specimens, is having an impact on endangered bird populations in Europe. Tourists love to buy foreign oddities, like turtle shells, rhinoceros horns, and even gorilla hands. They do not think that their purchases caused the death of an endangered creature. Ivory has been banned in recent years, and yet poaching of elephants continues because ivory is a popular carving material and makes for lovely decorations.

Perhaps the most controversial area of animal rights is that of vivisection, or animal experimentation. The controversy arises not so much from the animal deaths (which occur in far greater numbers in food consumption), but from the long-term and possibly tortuous manner in which the animals are killed. Modern science, for the last three centuries or so, has believed that animal testing is an accurate parallel to human testing, and so animals are used as substitutes. On the other hand, our litigious, lawsuit-happy society has increased animal testing also. Insurance companies require lots of animal tests of cosmetics and such products, so that they can deny any fault in future lawsuits, saying "we tested the product on X number of animals without ill effect." Another arena of animal testing is education. Medical doctors and veterinarians are trained in surgical techniques using living animals. One aspect that concerns the public is the source of such experimental animals. There have been reports of stolen pets being used in labs. However, some ask, is there any harm in taking animals from the "pound" or animal shelter, when they are about to be euthanized anyway? Furthermore, hundreds of thousands of animals are used in high school biology classes each year so that students can cut them open to see their internal anatomical structures. There are a growing number of alternatives to all of these practices, and some reductions in vivisection have been attained over the last decade. Many opponents have taken their objections into the marketplace, creating vegetarian food products and "cruelty-free" products, which have avoided any animal testing.

One fast growing area of controversy, new to the animal rights arena, is that of genetic engineering. Humans have altered animal genetics for millennia, but very slowly, through selective breeding and domestication. But our modern scientific techniques enable us to select certain genes and thus radically alter a species within a single generation. Will this lead to "Frankenstein" monsters, or improve our lives? Since we eat cows, would it be proper to engineer a cow without a brain or legs (that is fatter and less difficult to manage)?

Some animals are killed for religious reasons, as in sacrifices, but this is not so common as it was in centuries past.

Another major category of fatal uses for animals is the sporting arena. The most common sport killing is hunting. Legal hunting usually occurs during certain seasons of the year, with state licenses, and with the public understanding that the hunt is an attempt to cull excess population. Because civilizations have driven most predators to endangerment or extinction, "game" species have become overpopulated. Hunters say that they are simply keeping the animals from over eating their habitats and starving to death in the winter. In my part of the country, schools actually close at the beginning of hunting season; the sport is so popular that absences would make class pointless. However, opponents claim that the game animal populations are being kept artificially high and that "sport" is hardly a challenge for the human who has a gun, and the animal pays with its life for losing the game. Traditionally, hunting was considered to be a "man's sport," but in recent years women have become a growing part of the arena. Perhaps this is related to the increasing number of women who carry guns for self-defense purposes. What is the purpose of hunting? It is not food for sustenance, usually. Often it is for the antler rack to put on the wall as a trophy, or a social event for "male bonding" with friends. Illegal hunting is called poaching and is usually illegal either because the species being chased is endangered or because it is done out of season or in a forbidden area.

Fishing is another form of hunting. Some fishermen practice "catch and release," which means they do not keep or kill the fish; they catch it and let it go, just for the fun of the hunt. Others keep and eat the fish, or mount them as trophies. For a long time it was assumed that fish do not feel pain, but more recent studies question that assumption.

The question of animal euthanasia, for population control purposes, will be considered in the chapter on animal populations.

---

320. Adams, Carol J. *The Sexual Politics of Meat: A Feminist-Vegetarian Critical Theory.* New York: Continuum, 256 pp., 1991. (0826405134) "[H]ow the sexualization of animals and the animalization of women combines to form a dual oppression, maintained by a patriarchal foundation which expresses itself through meat as a metaphor of power." See [www.triroc.com/caroladams/]. Reviewed at [www.psyeta.org/sa/sa1.2/birke.html] and [www.ivu.org/books/reviews/sexual-politics-of-meat.html]. Feminism is a very strong influence in the animal rights movement. A majority of animal rights activists are single, middle-aged women. Many feminists see the animal rights movement as being parallel to the women's suffrage movement, in which women (now, animals) were controlled by men (now, our animal-using society) without voting choices of their own.

321. Allen, Tim, ed. *Animal Welfare Legislation, Regulations, & Guidelines: Bibliography, January 1990 — January 1995.* Upland, CA: DIANE Publishing Co., 57 pp., 1995. (0788119702) Read the whole work at [www.nal.usda.gov/awic/pubs/oldbib/qb9518.htm]. See a short history of the Animal Welfare Information Center (AWIC), which prints many bibliographic helps on animal rights issues, at [www.labanimal.com/col/reg11.html]. The author has written a number of bibliographies on animal issues for the National Agricultural Library, which are accessible on the Internet. This one contains 244 citations, not annotated.

322. American Medical Association (AMA). *Use of Animals in Biomedical Research: The Challenge and Response.* AMA White Paper series.

Chicago: AMA, 1988. Defends vivisection. See a related article at [www.nsplus.com/nsplus/insight/animal/labrats.html]. See a strongly anti-AMA website at [www.geocities.com/CollegePark/8273/ama.htm]. Traditionally the medical community has been in favor of animal experimentation. They claim that most of the miracles of modern medicine are directly attributable to such experiments.

323. Baird, Robert M., and Rosenbaum, Stuart E., eds. *Animal Experimentation: The Moral Issues*. Contemporary Issues series. Buffalo, NY: Prometheus, 182 pp., 1991. (0879756675) Sixteen essays of opposing views by doctors, lawyers, philosophers, ethicists, and psychiatrists, including one essay defending the Cartesian view. Contributors include Richard Ryder, Peter Singer, Tom Regan, Bernard Rollin, and Murry Cohen. Reviewed at [asa.calvin.edu/ASA/book_reviews/9-92.htm].

324. Baker, Robert M., et al., eds. *Animals and Science in the Twenty-First Century: New Technologies and Challenges*. Glen Osmond, Australia: ANZCCART, 129 pp., 1994. (0646224840) Based on a conference held in 1994. See the speakers and content list at [www.adelaide.edu.au/ANZCCART/AnimalScience.html].

325. Balls, Michael, et al., eds. *Animals and Alternatives in Toxicology: Present Status and Future Prospects*. New York: Wiley, 390 pp., 1997 (originally published in 1983). (047119929x) The Proceedings of a 1990 conference to discuss a report by The Fund for the Replacement of Animals in Medical Experiments (FRAME). Read an article about toxicity testing at [www.nsplus.com/nsplus/insight/animal/toxic.html]. The testing of chemicals on animals to check for toxicity has been a major part of modern vivisection. In 1999, Vice President Al Gore was under attack from the animal rights side for supporting a major increase in the number of toxicological tests to be performed by the Environmental Protection Agency (EPA). Usually the animal rights people are unified with the environmental people, in regards to the protection of forests and habitat; however, environmentalists are generally more concerned with the health of "the whole Earth" and not the individuals in it. The greater good is sometimes met by sacrificing the individual good, they believe.

326. Bateman, James A. *Animal Traps and Trapping*. Harrisburg, PA: Stackpole, 286 pp., 1988 (originally published in 1970). (0715353403) A thorough study of trap development and legislation affecting its use. Read a pro trapping article at [www.wildlifedamagecontrol.com/factsar.htm] and an anti trapping article at [www.fur.co.uk/furage.htm]. Trapping is a somewhat obsolete practice in the United States, at least in regard to the fur industry, which receives most of its pelts from fur farmers on fur

ranches. However, trapping is still employed in many areas for the snaring of "pest species" like coyotes. The most controversial point about traps, perhaps, is the lengthy captivity. Traps were generally designed to capture and hold an animal without killing it or harming its pelt; and since the traps may only be checked every few days, the animal may be in pain for a long time. Another problem with traps is that they capture indiscriminately; sometimes catching dogs and cats, or other non-pest species.

327. Bauston, Gene. *Battered Birds, Crated Herds: How We Treat the Animals We Eat.* Watkins Glen, NY: Farm Sanctuary, 64 pp., 1996. (0965637700) A brief summary of modern food production, known as factory farming. Includes a short chapter on fish production, which is neglected in many animal rights works. Includes photographs. See excerpts at [factory farming.nettinker.com/farmsanctuary/birdherd/index.htm]. Read an article about the author at [www.farmsanctuary.org/media/latimes2.htm]. Factory farming is one of the most frequently cited reasons for vegetarianism, since the conditions of animals' lives in such places are vastly different from those of the "family farm." Family farms have been largely supplanted by huge industrial complexes, where animals are a "crop."

328. Bennett, B. Taylor, Brown, M. J., and Schofield, J. C. *Essentials for Animal Research: A Primer for Research Personnel*, 2nd ed. Chicago: University of Illinois, 84 pp., 1990. (0788136356) Read the whole book at [www. primate.wisc.edu/pin/awa/essentia]. The proper training of animal research personnel is of great importance to modern laboratories, since such sites are subject to licensure and enforcement regulations, and also because the experiments themselves may be affected by unnecessary animal stress.

329. Benson, Verel W., and Witzig, Thomas J. *The Chicken Broiler Industry: Structure, Practices, and Costs.* Agricultural Economic Report #381. Washington, DC: Department of Agriculture, 53 pp., 1977. Read an extensive article at [www.mcspotlight.org/media/books/fawn.html]. The poultry industry is the factory farming institution most reviled by animal rights supporters because it is the most restrictive on the animals (even more restrictive than the veal calf environment) and by far the largest. Over years of selective breeding and genetic engineering, modern broiler chickens are "unnaturally" large, since sales are based upon the weight of the bird. Because crowded conditions make chickens more aggressive than usual, they often have their beaks and nails cut off so as to reduce injuries. Download an Adobe file with a short article about the chicken broiler industry at [searchpdf.adobe.com/proxies/2/60/44/2.html]. See an article on how such an operation is started (detailed instructions) at [www.msue. msu.edu/msue/imp/modpo/e7730003.html].

330.Bergman, Charles. *Orion's Legacy: A Cultural History of Man as Hunter*, reprint ed. New York: Dutton, 365 pp., 1997. (0452275598) A very interesting work by an English professor at Pacific Lutheran University. He traveled the world to meet hunters of all kinds, in order to study them. He discusses how men assert their masculine identities by killing animals. This is a scholarly, sociological sort of work. Read two articles by the author at [www.nwf.org/nwf/intlwild/snowmonk.htm] and [. . . /span wolf.html]. Why do men (and women) hunt? Is it a genetic compulsion, based on millennia of human regression to the days when hunting was survival? In the modern world, hunting is largely a sport, not a subsistence requirement. Many hunters speak of their wish to "get in touch with nature." How is the killing of an animal a connection to nature? Perhaps because it is an invigorating adventure that occurs outside of the normal, urban, work world.

331.Bixby, Donald E., and Christman, Carolyn J. *Taking Stock: The North American Livestock Census*. Granville, OH: McDonald & Woodward, 182 pp., 1994. (0939923351) "A compendium of the current status of livestock in North America, calling for changes in practice and policy that will provide for sustainable agriculture in the future. The focus of this book is the crisis of genetic erosion within the livestock species of North America." (advertisement) Related articles at [www.sciencenews.org/sn_arc97/10_4_97/bob1.html] and [www.fao.org/sd/EPdirect/EPre0042.htm]. The question of "genetic erosion" arises with almost all domesticated animals, and even sometimes among people. By always breeding the largest or most beautiful cows, we may also be spreading undesirable characteristics that will not be recognized until the future generations of the animals.

332.Blum, Deborah. *The Monkey Wars*. New York: Oxford University, 306 pp., 1994. (019510109x) The author won a Pulitzer Prize for the articles that were combined into this book (written for the *Sacramento Bee* newspaper). They are interviews with researchers and activists regarding research on primates. See a useful site including a long list of available video resources on primates and primate studies at the Wisconsin Regional Primate Research Center [www.primate.wisc.edu/pin/av.html]. One of the videos listed is a 70-minute videotape done by the WRPRC in 1994, interviewing Deborah Blum, called *The Monkey Wars*. Read an interview with the author at [arrs.envirolink.org/ar-voices/tranblum.html].

333.Botting, Jack H. *Animal Experimentation and the Future of Medical Research*. Proceedings series, volume 3. London: Portland Press, 107 pp., 1992. (1855780380) The author concludes that some animal research is necessary for medicine. See an article at [www.sciam.com/0297issue/0297botting.html]. After years of scorching debate, some scientists have been willing to grant that there is "too much" animal experimentation, and that such must be reduced to more minimal levels. However, many animal rights supporters hold an "all or nothing" position, which requires the end to all experiments, for any reason. Is there a middle ground? If so, it has not become popular with either side of the argument.

334. Bryan, Richard. *Why do People Fish?* Washington, DC: Sport Fishing Institute, 1974. Read an article about a survey done on Kentucky fishermen at [www.iglou.com/fishincom/articles/attitude.html]. This is a good question, similar to "why do people hunt?" Some people, who are largely vegetarian, continue to eat fish even after they have given up all other meats. Why? Fish tend to be viewed as "lower" than mammals and birds. We have far less contact with fish than with other creatures, since humans do not live in the watery environments. On the other hand, most people who own pet fish would probably be horrified at the idea of eating their pet fish. Why? Because as pets the fish have acquired a greater respect by the owner. It is also a common belief that fish cannot feel pain, which is a disputable idea.

335. Buyukmihci, Nedim C. *Alternatives to the Harmful Use of Nonhuman Animals in Veterinary Medical Education.* Vacaville, CA: Association of Veterinarians for Animal Rights, 1990. Read the whole work at [www.avar.org]; and see a similar article on the subject at [www.avar.org/surgery.htm]. In the last two decades, a growing number of alternatives to animal experimentation have been devised, with varying levels of success. Some use cultures of cells, which as living tissue can react to toxicity in similar ways to living creatures. There are improving computer models that simulate reactions based on other known data. Still, most scientists say that we are a long way from completely replacing animal experiments with such alternatives.

336. Cartmill, Matt. *A View to a Death in the Morning: Hunting and Nature Throughout History.* Cambridge: Harvard University, 331 pp., 1993. (067493735x) A critique of the anthropological theory that man has been a natural hunter throughout his long evolution. How hunting became popular. Reviewed at [www.psyeta.org/sa/sa4.1/marvin.html]. See a brief biography and bibliography of the author at [www.baa.duke.edu/FacPages/cartmill.html]. Many hunters believe that they come to the sport of stalking and killing animals naturally because it is "normal" for man to hunt and has been for millions of years of evolutionary history. On the other hand, why is it that 70 or 80 percent of the American populace does not feel this primal urge to hunt? If the human progression truly went from a hunting society to a hunter/gatherer society and then to the farm civilization, is hunting a vestigial (unnecessary) throwback to ancient history?

337. Christensen, John O. *Animal Experiments/Animal Rights: A Bibliography of Recent References*. Public Administration series. Monticello, IL: Vance Bibliographies, 1991.

338. Coats, C. David, and Fox, Michael W. *Old MacDonald's Factory Farm: The Myth of the Traditional Farm and the Shocking Truth about Animal Suffering in Today's Agribusiness*. New York: Continuum, 186 pp., 1989. (0826404391) See a related page at [www.veganoutreach.org/wv/wvan.html]. Factory farming is the most common method of meat production in modern times in first-world nations; family farms are becoming extinct. Yet our society still pretends that family farms are the norm. Why? Certainly because it is more pleasant to do so. Television commercials and billboard ads show lovely pastoral settings with cows grazing and chickens scratching, though in fact such scenes are completely inaccurate in regard to the truth of factory farming. Is this a lie, or just good marketing? And why have we gone to factory farms? Because Americans (and other first-world nations) have learned to eat meat three times a day, the demand for meat is so high that family farms could not maintain such a huge number of cattle and poultry.

339. Coe, Sue. *Dead Meat*. New York: Four Walls, 136 pp., 1996. (156858041x) Something of a sequel to Upton Sinclair's *The Jungle*, written around the turn of the century (see this entry below in this chapter). Coe spent six years travelling to various slaughterhouses but was not permitted to take photographs. So she did drawings instead, 125 very graphic and disturbing images of what she saw. She tells of the horrible waste, inefficiency, and haste in these places. Reviewed at [webdelsol.com/FLASHPOINT/suecoe.htm]. Isn't it strange that slaughter-houses do not permit photographs? Whatever happened to "freedom of the press"? Some scenes are considered to be too horrific for the public to view. Then you have to wonder why it is acceptable to tolerate such actions if the actions cannot even be viewed by the average person. In ancient times, unless you were rich and had slaves or servants, when you wanted to eat meat, you killed it yourself. These days, only a chosen few actually kill animals, then the public purchases a sanitized and packaged version for self-preparation.

340. Cohen, Murry J., and Natelson, Nina, eds. *The Future of Medical Research Without the Use of Animals: Facing the Challenge*. Alexandria, VA: Concern for Helping Animals in Israel (CHAI), 189 pp., 1990. (09631-59607) Thirteen articles taken from the First International Medical Conference in Israel. Includes articles by Gary Francione and Pietro Croce. Read articles by the author at [arrs.envirolink.org/chai/cohen.htm] and [www.mrmcmed.org/ape.html]. I have also seen this book listed with the title reversed: *Facing the Challenge: The Future of Medical Research*

without the Use of Animals. I read recently that pro vivisectionist scientists were able to threaten and pressure the Israeli academic institution (where the year 2000 CHAI conference was to be held) into canceling it.

341. Coleman, Vernon. *Why Animal Experiments Must Stop.* London: Green Print, 106 pp., 1995. (1854250604) This prolific medical doctor (author of nearly 100 books) writes on the subject of vivisection. See a similar book online by Coleman at [www.vernoncoleman.com/books/ffa.htm]. Read an article by the author [arrs.envirolink.org/ar-voice/how2win.html].

342. Commission of the European Communities. *Meat Consumption in the European Community.* Lanham: CEC, 215 pp., 1994. (9282670341) See a related article at [chianti.stat.unibo.it/fanfani/sintesiAIR.HTM]. We Americans tend to forget that the rest of the nations of the world have their own customs that may be different from our own. In Russia and Eastern Europe, most people only eat meat two or three times a month. Is it just because they are poor, compared to us? If they had more money, would they eat more meat?

343. Cook, Lori. *A Shopper's Guide to Cruelty-Free Products.* New York: Bantam, 262 pp., 1991. (0553292080) A listing of commercial products whose producing companies have announced that no animal testing is done in the creation of those products. See a list of charities that do not sponsor experiments [www.perm.org/issues/charities.html]. This consumer tactic of listing products whose manufacturers meet certain requirements (in this case, not testing the products on animals) has become a very effective tool for activists. Companies who are not on this list lose potential sales, which is always a motivating factor for change. Furthermore, it sets up a place for new and smaller companies to have a "niche" market, in providing products they know will sell. Capitalism is a good arena for this type of protest, where animal rights supporters have created a market for cruelty-free products.

344. Cooper, David K. C., and Lanza, Robert P. *Outwitting Evolution: Transplanting Animal Organs into Humans.* Oxford: Oxford University, 272 pp., 2000. (0195128338) This book has not yet been printed, but the announcement and title makes the book seem relevant to the animal rights debate (and the practice of xenotransplantation is always fatal to the animal). Xenotransplantation, the transfer of an organ from one species into a different species, is a new trend in surgery. It has been performed on a fairly limited basis so far, but as new pharmaceutical drugs make organ rejection less frequent, I believe it will become more common.

345. Covino, Joseph, Jr. *Lab Animal Abuse: Vivisection Exposed.* Berkeley, CA: New Humanity, 533 pp., 1990. (0943283000) Read an excerpt at [www.greenepa/~jtsai/trail/1997/cruelty.htm]. See an article about guilt among vivisection workers by Arnold Arluke [www.nsplus.com/nsplus/insight/animal/guilt.html]. One problem with vivisection labs is maintaining a balanced atmosphere of kindness with hard-heartedness. In other words, it is no easy thing to be friendly with an animal most of the time, and then to harden your heart to strap into a device and torture it with chemicals and knives. Human hearts are not designed well for this kind of contradiction. To protect themselves emotionally, many workers always act with disdain for the animals so as not to feel so guilty about the testing later. This is one of the difficulties of vivisection lab workers, and perhaps one reason why abuses are found.

346. Croce, Pietro. *Vivisection or Science: An Investigation into Testing Drugs and Safeguarding Health*, 2nd ed. New York: Zed Books, 1999 (originally published in 1981). (185649733x) The author (who spent his early years as a vivisectionist) says that vivisection is not a true science. Read the whole book at [www.pnc.com.au/~cafmr/online/research/croce1.html]. Read an article by the author at [www.linkny.com/~civitas/page9.html]. One issue in vivisection is the purpose motivating it. Much of modern animal experimentation has nothing to do with learning new things but simply has to do with protecting companies from lawsuits based on their products. Insurers demand a certain number of animal tests before covering companies from lawsuits. So nothing new is being learned; in reality, it is just that X number of rabbits must be injected with X product so that in a trial, the company can say "we tested that." The amazing number of frivolous lawsuits filed by consumers is one reason for high levels of animal experimentation.

347. Davies, Brian. *Red Ice: My Fight to Save the Seals.* London: Methuen, 347 pp., 1991 (originally published in 1989). (0750500611) Read a large article on seals and sealing at [www.american.edu/projects/mandala/TED/HARP.HTM]. The Canadian government claims that seal populations are actually climbing, and that the herds must be culled to protect the fish industry. Brian Davies's immense publicity campaigns against sealing in America and Europe led to major cuts in seal-fur buying and the inclusion of the harp seal in the 1972 Marine Mammals Protection Act, even though these seals are not endangered.

348. Davies, Brian. *Seal Song.* New York: Viking, 94 pp., 1978. (0670626686) A case for the preservation of seals. See a web page in favor of seal hunting at [www.naiaonline.org/seal1.html]. One question about the killing of baby seals for fur is the method used. In order to protect the pelt from any

damage, the seals are clubbed on the head. Activists say that many of the seals are only stunned, and thus are skinned alive. Supporters of the hunt say that clubbing is more effective even than shooting them. Another question is the immense amount of resources that the Canadian government has used to protect this business of sealing, which is hardly a wealth-producing industry and is only profitable for a few thousand people. The Royal Canadian Mounted Police seem to spend far more money arresting protesters than is worthwhile for the economics gained by the hunts.

349. Davis, Hank, and Balfour, Dianne, eds. *The Inevitable Bond: Examining Scientist-Animal Interactions*. Cambridge: Cambridge University, 399 pp., 1992. (0521405106) How human scientists and researchers become attached to their animal subjects, and how that attachment may affect the animals and the results of the experiments. This is not so much intended to stop necessary interaction but to minimize its impact on the tests. This is accomplished by redesigning the tests and environments in which the animals are kept. There is a good review at [www.newscientist.com/nsplus/insight/animal/uncertainty.html].

350. Davoudian, Hoorik. *Animal Experimentation: The Hidden Cause of Environmental Pollution? Absolutely!* Glendale, CA: SUPRESS, 10 pp., 1993. Read this booklet online at [members.xoom.com/vegan/vivpol.html]. Reviewed at [www.pnc.com.au/~cafmr/reviews1.html#animal]. The author is vice president of SUPRESS, which does not call itself an animal rights organization, but simply an anti-vivisection group.

351. Day, David. *The Whale War*. San Francisco: Sierra Club Books, 168 pp., 1987. (0871567784) "The whale is at the heart of a guerrilla war of resistance that has spread all over the world: it is the symbol of the ecology movement . . . If this amazing animal, the largest ever to exist on the planet, cannot be saved from the ruthless exploitation of a handful of men, what chance of survival have other species?" (p. 1). Long discussions of the effectiveness of Greenpeace and the International Whaling Commission (IWC). See the Greenpeace organization's reaction to Norway's continued whaling despite the IWC moratorium at [www.greenpeace usa.org/media/publications/norwhaletext.htm].

352. Day, Nancy Raines. *Animal Experimentation: Cruelty or Science?* Issues in Focus series. Springfield, NJ: Enslow, 128 pp., 1994. (0894905783) Written for young adults. See a webpage by high school students on this subject at [coleharbourhigh.dartmouth.ns.ca/975331/animal.htm].

353.DeDeyn, Peter P., ed. *The Ethics of Animal and Human Experimentation.* London: John Libbey, 1994. See a good article at [www.perm.org/issues/Animal_Experimentation_Issues/understanding_claims.html].

354.Diaz-Canabate, Antonio. *The Magic World of the Bullfighter.* London: Burke, 323 pp., 1956. For a list of books about bullfighting see [mundo-tautino.org/my_books.html]. Surprisingly, there are very few English-language books about bullfighting, perhaps because it is not a sport in the United States. It is certainly an animal rights issue, however. Where practiced, in places like Mexico and Spain, bullfights traditionally end in the public spectacle of the bull being stabbed to death with a sword. Proponents say that it is a fair fight, with the bull having a chance to kill the matador. Opponents say that the bulls are drugged and do not have a good chance in the contest. It is somewhat reminiscent of the ancient Roman arena games, where animal and human deaths were a spectator sport.

355.Dickinson, Lynda. *Victims of Vanity: Animal Testing of Cosmetics and Household Products and How to Stop It.* Toronto: Summerhill Press, 108 pp., 1989. (0920197906) This former model, who has become a vegetarian, writes against the testing of cosmetics and household products on animals. See related articles at [www.ufaw3.dircon.co.uk/Boydcosmetics.htm], [www.straightedge.com/pages/anti-ignorance/product.html], [www.Nonline.com/procon/html/conAT.htm] and [. . . proAT.htm]. Cosmetics testing is a large part of the animal experimentation market. Since cosmetics are often applied to the human face, where the eyes and mouth give quick entry into the human body, the contents of each product are tested on many animals to see if they are toxic or cause discomfort. Rabbits have been the traditional animal of choice for such testing, with the infamous Draize test. Because rabbits do not have the ability to blink eyelids, they are easily restrained, have their eyes filled with the cosmetics, and can be observed for irritation. For the most part, this is more companies' efforts to protect themselves from lawsuits by customers who may complain of irritation from a given product.

356.Diner, Jeff. *Experimental Psychology: Experiments Using Animals.* Clark's Summit, PA: International Society for Animal Rights, 1987. See the American Psychological Association guidelines at [www.apa.org/science/anguide.html]. The irony of the long-practiced psychological experiments on animals is that most scientists have denied that animals can feel pain. Isn't it strange to run psychological tests upon creatures that are so dissimilar to humans that they cannot even sense pain? Perhaps there could be some physiological data gleaned there, but psychological? In modern times, most scientists have come around to admit that animals do feel pain.

357.Dizard, Jan E. *Going Wild: Hunting, Animal Rights, and the Contested Meaning of Nature*, revised and expanded ed. Amherst: University of Massachusetts, 232 pp., 1997 (originally published in 1994). (0870239082) The author uses a recent controversy over deer hunting, at the Quabbin Reservoir in Massachusetts, to get a handle on some of the issues and feelings of involved parties. After interviewing animal rights activists, hunters, park managers, and environmentalists, the author represents the groups by categorizing them as holding two different worldviews: those who think that nature is fragile and should not be tampered with, and those who think that nature is dangerous and requires human intervention and management. The author has very astutely labeled the two parties the "conservationists," believing that nature needs our help to survive, and the "preservationists," who believe that nature should be protected from human intrusions. These two worldviews shape and color many environmental and animal rights debates.

358.Dolan, Kevin. *Ethics, Animals and Science.* Ames: Iowa State University Press, 320 pp., 1999. (0632052775) The publisher advertisement says that the book is written for laboratory personnel and offers many different viewpoints on the ethics of animal experimentation, without drawing any polemical conclusions. The same author is scheduled to publish an *Introduction to Laboratory Animal Law* in the year 2000. See the publisher advertisement, including the Table of Contents, at [store.yahoo.com/isupress/0632052775.html].

359.Doncaster, Anne, ed. *Skinned: Activists Condemn the Horrors of the Fur Trade.* North Falmouth, MA: International Wildlife Coalition, 256 pp., 1988. (0685261093) Link to a 13 minute radio interview with the author at [webcom.com/iwcwww/IssuesProjects/SealHunt/anne.html]. See the publisher's website with links at [http://www.iwc.org/]. The fur trade, though greatly reduced in size and popularity by cultural changes (and animal rights protests, perhaps), continues to be a major target of activists. In fact, "direct" and illegal actions have increased in recent years, with the Animal Liberation Front (and similar groups) breaking in to fur farms and freeing the fur-bearing animals. One problem with this tactic is that most of the animals tend to die. Fur farm animals have never lived in the wild and do not know how to feed themselves, hide from predators, and so on. Many are killed by cars on roads or by local dogs and cats. Nevertheless, the ALF says that their purpose is to put fur farms out of business, which these actions may in fact do, while yet killing the creatures themselves.

360.Doughty, Robin W. *Feather Fashions and Bird Preservation: A Study in Nature Protection*. Berkeley: University of California, 184 pp., 1975. (0520025881) This may have been the world's first widespread conservation movement, led by wealthy women who did not want fashion to destroy the many beautiful bird species. For a long time, it was considered to be the height of fashion for women to wear huge hats full of flowers and bird plumage. Several species of birds became endangered or even extinct due to this fashion. A similar problem in recent years is that of egg collecting. Some ornithologists become obsessed with having a sample egg of every possible bird species, and thus have endangered bird species all over Great Britain. A safer hobby (for the birds) would be simply to photograph the egg or bird, rather than claim and kill it. See the author's webpage at [www.utexas.edu/depts/grg/office/faculty/faculty/doughty.html], and his bibliography at [http://www-geography.berkeley.edu/Publications/60YrsGeog/Doughty,%20Robin.html].

361.D'Silva, Joyce, and Tansey, Geoff, ed. *The Meat Business: Devouring a Hungry Planet*, revised ed. New York: St. Martin's, 256 pp., 1999 (originally published in 1994). (0312226861) The author is the director of Compassion in World Farming; see a brief biography at [ivu.org/evu/eurocong99/joyce.html]. Read an excerpt at [www.greenepa/~jtsai/trail/1997/cruelty. htm].

362.Eaton, Randall L. *The Sacred Hunt (parts 1 and 2): Hunting as a Sacred Path* (book and video). Enterprise, OR: Sacred Press, 208 pp., 1998. (0966369610) These are very strongly pro-hunting. A number of different reasons and philosophies are offered in support of the practice of human sport hunting of animals, including Native American and religious viewpoints. See the author page at [www.eoni.com/~eaton]. Reviewed at [www.mailtribune.com/outdoor/archive/101697o2.htm]. Music for the videos was written by Ted Nugent, an outspoken hunter and gun advocate. The videos are 90 minutes (part 1) and 60 minutes (part 2) long, and have a fairly high production appearance.

363.Eisnitz, Gail A. *Slaughterhouse: The Shocking Story of Greed, Neglect and Inhumane Treatment Inside the U.S. Meat Industry*. Buffalo, NY: Prometheus, 310 pp., 1997. (1573921661) A chief investigator for the Humane Farming Association tells of the problems (observed in a ten-year investigation) in slaughterhouses. Foodborne illnesses, foul meat, injuries due to production line speeds, dismemberment of animals while alive, non existent enforcement of humane laws, human worker injuries and stress, and USDA complicity to overlook meat industry violations. Read a substantial review at [www.hfa.org/slaughterhouse.html]. Read an interview with the author at [www.montelis.com/satya/backissues/jan98/

farming.html]. See articles at [arrs.envirolink.org/ar-voices/a_slaughter house.html] and [. . . /slaughter.html]. See an article promoting animal welfare in the slaughterhouses at [lamar.colostate.edu/~grandin/welfare/ general.session. html]. Aside from the simple issue of the horrors perpetrated on animals, this book brings up the parallel issue of the effects upon the workers. In the not-so-distant past, the British legal system would not allow butchers to serve on juries because they were considered to have become incapable of sympathy for sufferers, human or animal. What is the impact on a person's "humanity" to kill hundreds of animals per day in a slaughterhouse? This sounds like a sociology study waiting to be done.

364. Emberley, Julia V. *The Cultural Politics of Fur.* Ithaca, NY: Cornell University, 272 pp., 1998. (0801484049) The author teaches Women's Studies and Gender Studies in Canada. How fur is a symbol of wealth, sexuality, and social class. Feminist Carol Adams writes a fairly critical review of this book, which you can read at [mosaic.echonyc.com/~onissues/ sp98fur.html]. Is the wearing of fur simply a symbol? Certainly it is no longer necessary, with the readily available alternatives of cotton and synthetic fibers. How about fake furs? What is the purpose of wearing a fake fur? Is it simply attractive? Or does it make a statement?

365. Evans, E. P. *The Criminal Prosecution and Capital Punishment of Animals: The Lost History of Europe's Animal Trials*, 2nd ed. reprint. Union, NJ: Lawbook Exchange, 384 pp., 1998 (originally published in 1906). (1886363528) Reviewed at [www.psyeta.org/sa/sa2.1/beirne.html]. See a related page at [www.aarrgghh.com/no_way/clawClaw.htm]. Strangely, animals have been tried, convicted, and executed for crimes against human law.

366. Fadali, Moneim A. *Animal Experimentation: A Harvest of Shame.* Los Angeles: Hidden Springs, 288 pp., 1997. (1885113552) A cardiovascular surgeon uses scientific evidence to show that animal research is unnecessary. See the table of contents and several short reviews at [www.bookzone.com/bookzone/10001130.peek.html]. Read an article by the author and chapter one of the book at [www.geocities.com/College Park/8273/fadali.htm]. Assuming that the author is correct, that animal research is not necessary, does it also follow that animal research is not desirable? Perhaps alternative methods are somewhat reliable, but say (for the sake of argument), they are only 90 percent as reliable as true animal experimentation. So do we go with the 90 percent reliability rather than the 99 percent (random figures)? If there is some benefit to humans from vivisection, do we abandon that small benefit?

367.Fano, Alix. *Lethal Laws: Animal Testing, Human Health and Environmental Policy*. New York: St. Martin's, 242 pp., 1998. (1856494977) The author is coordinator of the Medical Research Modernization Committee. Why animals are not needed in toxicological research. "[This book] effectively demonstrates the causal link between animal testing and environmental degradation and the subsequent deterioration of human health." See the author's page at [alixfano.home.mindspring.com]. Extensive reviews can be found at [www.linkny.com/~civitas/page43. html] and [www.montelis.com/satya/backissues/feb98/views.html]. This is an interesting twist. Rather than saying that animal research is unneeded, or that other methods are available, this author says that animal experimentation is actually causing human harm.

368.Fiddes, Nick. *Meat: A Natural Symbol*. London: Routledge, 250 pp., 1992. (0415089298) Why is meat so popular? There was a major increase in meat consumption during the Industrial Revolution, due perhaps to feelings of control over nature. The author is not a vegetarian himself, but he asks why meat-eaters are so hostile to vegetarians. Reviewed at [www. psyeta.org/sa/sa1.2/birke.html]. An interesting question: why are meat-eating folks so defensive when they meet vegetarians? Is it guilt, or pride, or something else? One radio commentator often says "we humans didn't fight our way to the top of the food chain to eat tofu burgers!" Quite humorous, but why is the satire needed? It leads one to wonder whether meat-eating is not something of an addiction; since this reaction is similar to that of a cigarette smoker or an alcoholic when challenged with quitting.

369.Fox, Michael Allen. *The Case for Animal Experimentation: An Evolutionary and Ethical Perspective*. Berkeley: University of California, 262 pp., 1986. (0520060237) The author later published retractions of this work. "This book is an essay in support of the use of animals for human ends. Not just any animals and not just any ends, and most important of all, not without important qualifications." (p. 1) Briefly reviewed at [www.netsurf.com/nsd/books/book.04.02.html]. Animal rights supporters may gloat because this author later changed his mind and disowned this book. Nevertheless, it is considered to be a classic on the pro vivisection side, and his preconversion ideas should be considered as concepts worthy of thought.

370.Fox, Michael Allen. *Deep Vegetarianism*. America in Transition series. Philadelphia: Temple University, 224 pp., 1999. (1566397049) This once outspoken pro vivisectionist is now an animal rights activist (see the entry immediately above). See the author's page at [qsilver.queensu.ca/philo sophy/faculty/Fox.htm]. One observation: people do sometimes change. Activists on both sides can become frustrated and think that no one is listening to what they have to say. Here is a case where the author was listening and changed his mind (rightly or wrongly).

371.Fox, Michael W. *Eating with Conscience: The Bioethics of Food.* Troutdale, OR: New Sage Press, 224 pp., 1997. (0930165309) How factory farms destroy the environment and our health. How consumers can reshape our food system. One chapter on genetic engineering, and one on fish farms and overfishing. Briefly reviewed at [www.netsurf.com/nsd/books/book.04.02.html]. See an article at [www.teleport.com/~newsage/books/eating_prev.html]. It is fascinating how rarely we humans consider the ethics of our actions. After all, most Americans do eat animals two or three times a day. How many times have those Americans stopped to consider whether this was a good or bad thing? It just is, because everybody is doing it? Humans are very traditional, and tend to do whatever their parents did.

372.Fox, Nicols. *Spoiled: The Dangerous Truth about a Food Chain Gone Haywire.* New York: Penguin, 224 pp., 1998. (014027555x) About the dangers of bacteria and disease in modern meats. Read an interview with the author at [www.emagazine.com/may-june_1998/0598conversations.html]. Is food poisoning becoming more common? Or is the media simply making the cases more public?

373.Francione, Gary L., and Charlton, Anna E. *Vivisection and Dissection in the Classroom: A Guide to Conscientious Objection.* Jenkintown, PA: American Anti-Vivisection Society, 136 pp., 1992. (1881699005) Legal citations of relevant cases; sample letters to send to schools when asked to perform vivisection. Read an article about dissection training at [www.psyeta.org/sa/ sa2.1/lock.html]. See two articles of success stories in avoiding dissection in schools at [werple.net.au/~antiviv/issue169.htm] and [www.animalsalliance.ca/factsheets/dissect.html]. Most schools are willing to find alternative assignments for students who do not wish to participate in the cutting up of an animal in class. It is good to encourage children to take a stand on principles, if that is what they believe. If the point in resisting the dissection is simply to get out of work, don't bother. Usually the alternate work is harder than going along with the rest and doing the surgery!

374.Frazza, Al. *Animal Experimentation & Human Medicine.* People for Reason in Science & Medicine, 52 pp., 1995. (1886605009) Read the work at [www.livelinks.com/sumeria/ health/prism.html]. Most scientists say that animal experiments have led to longer human lives, due to the development of medicines. If true (this is hotly debated), does that make it right? Humans do see an ethical dimension in societal actions. Does the good end (better medicines) justify the means (animal vivisection)? If so, why not do such experiments on criminals, or "retarded" people?

375. French, Roger Kenneth. *Dissection and Vivisection in the European Renaissance*. The History of Medicine in Context series. Brookfield, VT: Ashgate Publishing, 304 pp., 1999. (1859283616) Read an article on dissection at [www. psyeta.org/sa/sa1.1/bowd.html]. It is never a waste of time to consider the history of an idea, no matter how ancient. At least, it shows how our society has come to be where it now is. The Renaissance was the rebirth of the sciences, long neglected, and has led to our modern world. See a brief publisher's description of this book at [www.ashgate.com/index2.cfm?visitingfrom=Americas].

376. Frey, Raymond Gillespie. *Rights, Killing and Suffering: Moral Vegetarianism and Applied Ethics*. Oxford: Blackwell, 256 pp., 1983. (0631126848) This philosophy professor (at Bowling Green State University) analyzes the three main moral arguments used for promoting moral vegetarianism and shoots them down. He says that animals do not have rights. See a brief faculty biography of the author at [www.bgsu.edu/offices/sppc/fellows.htm].

377. Friedman, Ruth, compiler. *Animal Experimentation and Animal Rights*. Oryx Science Bibliographies, volume 9. Phoenix, AZ: Oryx, 75 pp., 1987. (0897743776) A brief annotated bibliography with 245 entries of popular and recent works. See similar links at [animalrights.miningco.com/msubviv.htm].

378. Garattini, Silvio, and Bekkum, D. W. van, eds. *The Importance of Animal Experimentation for Safety and Biomedical Research*. Boston: Kluwer Academic, 1990. (0792305140) Read several articles in favor of vivisection at [www.ampef.org/articles/art_frame.htm]. See a FAQ (frequently asked questions) list at [www.aalas.org/fbr_faq.htm]. A number of books have come out in recent years defending the need for animal experimentation. This is good, whether their conclusions are agreeable or not, because it means that people are beginning to think about what they are doing. This is a big step in the right direction, from the days when scientists acted without considering the ethics of their actions.

379. Garrett, Tom. *Alternative Traps*. Washington, DC: Animal Welfare Institute, 61 pp., 1995. Read the table of contents and foreword at [www.animalwelfare.com/pubs/bk-conts.htm#traps]. See links at [www.animalwelfare.com/trapping/index.html]. A couple of books have now been published that offer "humane" methods of trapping troublesome animals, without harming or killing them. This does lead inevitably to the

additional work of transporting the captured creature to another location; but many people prefer that to ending its life.

380. Gaunt, Stanley N., and Harrington, Roger M., eds. *Raising Veal Calves*. Manual #106. Massachusetts Cooperative Extension Service, 59 pp., 1974. Read about an innovative new attempt to bring more "humanity" to the veal industry at [www.agwayagproducts.com/cooperator/april99/meadow veal.html]. See a Canadian pro-veal page at [www.ofac.org/veal.html]. Unfortunately, this is a neglected topic, from the industry side. Veal calves are generally male calves born to milk farmers. Of course male cows are not much needed at the milk factory. So the males are sold to veal farmers, who butcher them quite young. The main controversy comes from the methods of controlling and feeding the veal calves. The animals are in tiny pens, unable to walk or turn; and they are fed a strict diet to keep them anemic and weak. The reasoning is that the public likes their veal steaks soft and white, whereas any muscle would discolor the meat and make it tougher.

381. Gay, William I., ed. *Health Benefits of Animal Research*. Washington, DC: Foundation for Biomedical Research, 1987. The author was a "pioneer" in helping to set standards by which animal care could be measured, in the late 1950s and early 1960s. Pro vivisection people often say that animal rights supporters shouldn't talk about banning such experiments because "without those experiments you might not be alive today." While this may be true, the animal rights people themselves are not responsible for what was done by society before their own existence, and so it is not hypocrisy for them to oppose such work. As long as the animal rights people accept the fact that their own health may be affected by a ban on such work in the future, there is no hypocrisy in it. See a brief list of related articles at [www.orbiomed.org/index.html]. See an interesting perspective against animal research from the angle of determining Nobel Prize winners whose work did (or did not) make wondrous health advances against disease by using animal research at [www.aavs.org/Docs/nobel.htm].

382. Gay, William I., and Heavner, James E., ed. *Methods of Animal Experimentation*. Orlando: Academic, 382 pp., 1989. (0122780094) Seven volumes published between 1965 and 1989. Use great links to laboratory animal associations (in favor of vivisection) at [www.aslap.org/links.htm].

383. Gellatley, Juliet. *The Silent Ark*. San Francisco: Thorsons, 241 pp., 1996. (0722531621) Claims to offer the truth about meat farming, how and where the animals are produced for the industry, and the crimes committed in the name of profit. Read the introduction at [www.viva.org.uk/Silent Ark/silentark2.htm]. Reviewed at [www.ivu.org/evu/news962/book.html].

Is it wrong for industry to provide what the public demands? After all, that meat is all being sold. Or is the industry marketing its product to increase the demand? Is that wrong? Blaming the industry is somewhat justified, but the consumers who have created this vast market are equally to blame (if blame is due).

384. Gerstell, Richard. *The Steel Trap in North America: The Illustrated Story from Its Colorful Past to the Present Controversy*. Harrisburg, PA: Stackpole, 352 pp., 1985. (0811716988) The author worked for the Pennsylvania Game Commission. See a store that sells traps, with photographs, at [www.snareshop.com]. The practice of trapping is largely criticized for two reasons: (1) traps catch indiscriminately — pets, domestic animals, and people, as well as their intended victims, furbearers and predators; and (2) traps may cause lingering suffering and death, depending on how often the owner checks those traps. See a list of books at Amazon.com on trapping at [www.t3d.com/key_biological_physical_concepts/trapping.html].

385. Gluckstein, Fritz P. *Laboratory Animal Welfare*. Current Bibliographies in Medicine, No. 89-15. Washington, DC: Government Printing Office, 22 pp., 1990. See this 230-citation bibliography at [www.nlm.nih.gov/pubs/cbm/labanim.html]. The author works for the National Library of Medicine.

386. Gold, Mark. *Assault and Battery: What Factory Farming Means for Humans and Animals*. London: Pluto, 172 pp., 1983. (0861047273) Attacks the five myths that are used to defend factory farming. The author is one of Britain's most experienced campaigners and the former director of Animal Aid. See links at [www.animalwelfare.com/farm/index.html].) Factory farming is based on the modern plant-farming methods, where the most possible crop is grown in the least possible space, using the least possible human labor, and whatever necessary chemicals are required to maintain relative health for the product. A high number of expensive chemicals are required to keep the animals from catching diseases or injuries in the high stress, crowded environments; yet it is still profitable because of the reduced space and labor needs.

387. Gregory, Neville G., and Grandin, Temple. *Animal Welfare and Meat Science*. Wallingford, England: CAB International, 304 pp., 1998. (08519-9296x) Grandin is a professor of animal science at Colorado State University. View many of her articles at [grandin.com]. Read an article by Gregory at [www.maf.govt.nz/MAFnet/publications/9716/9716.html]. See a publisher's review and contents summary at [www.pighealth.com/

MEDIA/P/BOOKS/BKDETAIL/GREGORY.HTM]. How meat quality is related to animal welfare.

388. Groves, Julian McAllister. *Hearts and Minds: The Controversy over Laboratory Animals.* Animals, Culture, and Society series. Philadelphia: Temple University, 226 pp., 1997. (1566394767) Interviews both sides for their ideas and feelings. This is a study of "people's intellectual and psychological explanations for their positions." (backcover) Read an article by the author at [www.psyeta.org/sa/sa2.2/groves.html]. How do different people come to their different opinions? Some people adopt their parents' views, others adopt the media views, and others take views simply to rebel against the common views. Motive is never an easy question. Most people seem to take an opinion and stick with it while some folks do change or modify their opinions over time. Maturity seems to bring more change, perhaps because with a stronger self-image there is less discomfort or fear in consideration of new ideas.

389. Gudmundsson, Magnus. *Reclaiming Paradise?* (video). Reykjavik, Iceland: Megafilm, 1987. A pro sealing video meant to counter animal rights promoters. Quite critical of Greenpeace. Read two of Gudmundsson's articles online at [www.allianceforamerica.org/0596009.htm] and [www.allianceforamerica.org/issue2.htm].

390. Gudmundsson, Magnus. *Survival in the High North* (video). Reykjavik, Iceland: Megafilm, 1989 (originally released in 1986). A documentary on how natives survive by fur and subsistence hunting in Greenland, Iceland, and Forislands. It also shows evidence that at least one Greenpeace film released to the media faked a gruesome kangaroo slaughter in Australia in the early 1980s (which might imply a similar fake slaughter of seals in the seal hunt). Greenpeace has sued this independent filmmaker. One of the more difficult conflicts of the animal rights issue is that of subsistence hunting (hunting by local peoples for food and economic sustenance). It is, after all, rather difficult for Eskimos (Inuits) to survive in the Arctic without animal products since that is what they often eat and wear for clothing. On the other hand, it is also harder to feel sympathy for the natives when they take up modern weapons (rifles, speedboats, explosive harpoons) and sell the products in a commercial manner. See articles about this kind of human culture in the Arctic north at [www.grida.no/amap/assess/soaer5.htm] and [tceplus.com/nparctic.htm].

391. Guillermo, Kathy Snow. *Monkey Business: The Disturbing Case that Launched the Animal Rights Movement.* Washington, DC: National Press, 254 pp., 1993. (1882605047) Alex Pacheco, member of PeTA (People for the Ethical Treatment of Animals), infiltrated Maryland's Institute for

Behavioral Research, and took photographs and documented practices on monkeys. This led to the first cruelty charges brought against researchers. The Institute claims that Pacheco exaggerated and set up false exhibits for investigators. Why are the labs so secretive? A desire for secrecy does not guarantee that illicit activities are being done, but it does lead to suspicion. Is it that the general public does not wish to know what is going inside? Perhaps. The few charges that have even been brought against labs have been based upon inside information: either by former workers "blowing the whistle" or by animal rights "spies" who have infiltrated the facilities illegally or under false pretenses. Certainly, not all labs are doing illegal things. Some would argue that they all do immoral things, but even if true, immoral is not always illegal. See basic articles about PeTA at [www.the-scientist.lib.upenn.edu/yr1990/sept/opin1_900903.html] and [www.cyber soup.com/auction/charities/peta/petahistory.html].

392. Gunn, Brian. *Entering the Gates of Hell: Laboratory Cruelty You Were Not Meant to See*. St. Albans, PA: International Association Against Painful Experiments on Animals, 1987. Photographic scenes in laboratories. See an article about pain in laboratories at [www.animalwelfare.com/lab_animals/pratt/prat-idx.htm]. Photographs from lab experiments are rare; they are usually taken by "whistleblowers" or spies who wish to show the public what is going on.

393. Halverson, Diane, ed. *Factory Farming: The Experiment That Failed*. Washington, DC: Animal Welfare Institute, 86 pp., 1987. (0938414127) A compilation of Animal Welfare Institute articles on intensive farming and alternatives. Read excerpts at [www.animalwelfare.com/farm/factory.htm#book]. Many animal rights activists claim that factory farming is not only morally bankrupt but also financially inept. They say that without government subsidies and tax relief, factory farms would not be profitable ventures. The immense amount of antibiotics and drugs, and the mechanical devices used to raise the animals are quite costly. In order to cut costs, the farmers use cheaper and less healthy feed but then must increase drugs to combat the health consequences of such actions. Still, the number of factory farms has grown, while family farms have become fewer; so the profitability of factory farms seems to be adequate.

394. Harris, Marvin. *Good to Eat: Riddles of Food and Culture*. Prospect Heights, IL: Waveland Press, 289 pp., 1998 (originally published in 1985). (15776601512) Interesting information and links about the author at [www.voicenet.com/~nancymc/marvinharris.html]. This book was also published under the name *The Sacred Cow and the Abominable Pig*. How do people decide what kind of food is good to eat, and what kind is bad? Different cultures have wholly different views. Some cultures enjoy the

eating of several kinds of insects, while Western cultures eat none. Asian peoples enjoy dog meat and raw fish, which are usually odious to Westerners. Even in local regions of the United States, some foods are adored, like grits in the south, while in other regions people have no clue what grits are.

395. Harrison, Ruth. *Animal Machines: The New Factory Farming Industry.* New York: Ballantine, 215 pp., 1966 (originally published in 1964). This was the first detailed critique of factory farming. The author lists dangers to human health from these methods of production. Concern from this book led the British government to form "the Brambell Committee" (whose recommendations were promptly ignored by the bureaucracy). The foreword was written by the famed environmentalist Rachel Carson. Read a good article on factory farming at [www.amps.org/ru/arttreat.html].

396. Hart, Lynette A., ed. *Responsible Conduct with Animals in Research.* New York: Oxford University, 208 pp., 1998. (0195105125) The author teaches at the University of California, Davis. This is a basic overview of many issues in animal experimentation. Of course, the title implies an "animal welfare" approach, of making experiments less painful and less frequent; while "animal rightists" would say there should be no animals in research, the pain and frequency being irrelevant. See a good site with links for the European practices of animal experimentation at [www.doh.ie/policy/animal/].

397. Hendriksen, C.F.M., and Koeter, H.W.B.M., eds. *Animals in Biomedical Research: Replacement, Reduction, and Refinement: Present Possibilities and Future Prospects.* New York: Elsevier, 281 pp., 1991. (0444814175) These are the so-called Three R's of modern vivisection: replacement, reduction, and refinement. These three goals are to guide animal experiments in the future. See an article on the three R's at [www.jhsph.edu/~caat/pubs/articles/3r.htm]. They hope to replace as many such experiments as possible with noninvasive procedures; reduce the number of animal tests; and refine the tests to cause less pain and yet give good results. Critics say that such "improvements" more deeply entrench an evil system since they can now claim to "care." There also is some question as to whether any reductions have occurred. The vivisection industry is still quite large, particularly as pharmaceutical companies grows (and as the human population ages).

398. Henke, Janice Scott. *Seal Wars! An American Viewpoint.* Newfoundland: Breakwater, 215 pp., 1985. (091951961x) Henke says that the seal campaigns (to keep the baby Harp Seals from being clubbed for their furs) hurt the seals and the local residents. The first half of the book is detailed

scientific information about seals and seal hunting methods. She concludes that seals do not feel pain and are not social, intelligent, or emotional. The second half of the book is summaries of 'lies and evils' perpetrated by animal rights groups around the world. See an anti sealing article at [arrs.envirolink.org/ar-voices/seal_background.html]. Here is a conversion from animal rights to "conservation." The author says that we must kill more seals, not less, to save the fishing industries, which are collapsing. She also says that the spray paint used by activists to protect seals from clubbers hurts the seals by ruining their oily fur and making them cold. If the seal could understand and make a choice, however, I would think that it would take the paint rather than the club to the skull and skin removal.

399.Hepner, Lisa Ann. *Animals in Education: The Facts, Issues, and Implications*. Albuquerque, NM: Richmond Publishing, 305 pp., 1994. (09639-41801) A history of the use of animals in classrooms for dissection; the legal issues involved. This work was written by a student to help other students find alternatives to dissections in high school and college. See a related article at [www.nsplus.com/nsplus/insight/animal/dissecting.html]. High school biology classes are not the only places where this sort of activity occurs. Doctors-in-training and surgical veterinarians have often learned to do their cutting by practicing on living, anaesthetized dog-pound animals. These are animals doomed to euthanasia, put to use by the medical profession. Is this wrong? The animal will soon be dead anyway, and it is better to practice on a doomed dog than to make a surgeon's first attempt at surgery on a human, isn't it? Would you want to be the first person the surgeon cuts open? Does the end justify the means?

400.Hobbs, Claude A. *Indian Hunting and Fishing Rights II*. Washington, DC: George Washington Law Review, 1969. See a related article at [www.uhuh.com/laws/indhunt.htm]. The rights of native peoples have become a big issue again, particularly regarding hunting and fishing rights. Just this year, the Makah tribe of Washington state was granted the right to kill a few whales, as renewing its history and "rituals" of the past. This greatly complicates the United States' general opposition to whaling in the International Whaling Commission (IWC). Now Japan, Russia, and other whaling nations say "see, you are doing it too!" On the other hand, many of the old treaties that the U.S. government signed with these tribes say that they can hunt and fish on their own lands and local waters. Furthermore, the California grey whale is no longer considered to be an endangered species. See one article on modern Indian hunting as a possible problem for elk populations in the Pacific Northwest at [www.seattletimes.com/news/local/html98/hunt_032098.html].

401.Hobusch, Erich. *Fair Game: A History of Hunting, Shooting and Animal Conservation*. New York: Arco, 280 pp., 1980. (0668051019) This is considered to be a classic by hunters. Hunters are quick to point out that they were among the earliest conservationists, insisting that land be set aside for animals to live in and supporting the parks. This is true. Also, most hunters are not rude and uncaring poachers. They have no wish to see animal species go extinct. They may agree with bans on the hunting of some endangered creatures. But many animal rights activists demand an end to all hunting, which puts them at odds with sport hunters. See a brief article about hunting at [encarta.msn.com/index/conciseindex/60/060DC000.htm].

402.Horwitz, Richard P. *Hog Ties: The Role of Pigs and Pig Farming in American Culture*. New York: St. Martin's, 312 pp., 1992. (031221443x) Written by a midwestern professor of American Studies, as he remembers growing up on a pig farm and seeing the early stages of factory farms that were taking over the family farms. Good links on hog farming can be found at [www.animalwelfare.com/farm/index.html] Hogs are generally raised in the midwestern and southeastern United States, often in conjunction with nearby farmlands, where crops are readily available for their food. Factory farming of pigs came shortly after that of poultry. Interestingly enough, ethologists (animal behavior scientists) have concluded that pigs are among the most intelligent mammals. Many people have made pot-bellied pigs into beloved pets; pigs do not love mud unless they are too hot.

403.Houston, Pam, ed. *Women on Hunting*. Hopewell, NJ: Ecco, 336 pp., 1995. (0880014431) An anthology of 47 women writing pro and con on hunting, including fiction, poetry, and essays. Margaret Atwood is one of the authors. See related articles at [www.alloutdoors.com/alloutdoors/libhunt women.html] and [www.fund.org/facts/women.html]. This is an uncommon work because it is one of the few books to share female, non feminist views of an animal rights issue. The media does not often share the fact that women are increasingly buying firearms (usually for self-protection, not so much for hunting). Whether this trend is good or bad is up for debate; but it does show that women are not to be stereotypically grouped as wholly united for animal rights. Many women are hunters, vivisectors, and anti-animal rights activists.

404.Hummel, Richard. *Hunting and Fishing for Sport: Commerce, Controversy and Popular Culture*. Bowling Green, OH: Bowling Green State University, 186 pp., 1994. (0879726466) Discussions of the views of hunters and animal rights activists; new technological innovations that have changed hunting and fishing; and how movies and books portray the sports. A brief publisher review can be found at [www.bgsu.edu/colleges/library/press/pp0012.html]. Is "sport" a good enough reason to kill an animal or fish? The need for food is one thing; it may be debated as to whether meat is necessary, but certainly food and protein are human requirements. Sport is simply fun, however. And the fun is entirely on one side; the animal being chased is certainly not experiencing this fun. Some men see hunting as sort of a "rite of passage," a step toward true manhood. There may be practical reasons, as in overpopulation of deer, for hunting; but what is the purpose of sport hunting? In some ways it is a social event, male-bonding of a few men sharing a common goal.

405. Inglis, Leslie R. *Diet for a Gentle World: Eating with Conscience*. Garden City: Avery, 144 pp., 1993. (0895295814) "[A]nimal-based food consumption is truly hazardous to our health . . . . [and] under-publicized devastation being inflicted on our environment." This book discusses factory farming. See a related article at [www.dir.ucar.edu/esig/dale_animal. html] Some people are vegetarians just because they do not like the taste or health of meat products but many vegetarians now abstain from meat due to moral considerations.

406. Inglis, Leslie R. *Life After Beef: Reflections of a Vegetarian*. Florida: Cabbage Palm Press, 112 pp., 1992. (0963149601) Autobiographical description of Inglis's change from meat-eater to vegan. Find vegetarian links at [www.veg.org/veg]. A vegetarian is a person who abstains from eating meat, although some people call themselves vegetarians while continuing to eat fish, which they do not think of as "meat." A vegan goes a step further. Believing that any use of any animal product is wrong, they will not drink milk, eat eggs, or wear leather. Becoming a vegetarian is more difficult than you might think. Restaurants are full of meat products, it is not easy to find a pleasing variety of dishes that do not include meat. Becoming a vegan would be even more difficult. Dropping out all milk and cheese products is a major step.

407. International Whaling Commission. *Cetacean Behavior, Intelligence and the Ethics of Killing Cetaceans*. Washington, DC: Smithsonian Institution, 1980. Information on the International Whaling Commission at [www1.pos.to/~luna/whale/iwc.html]. See many articles and links at [www.animalwelfare.com/whales/indexout.html]. The International Whaling Commission is not popular with anyone, as most "moderate" groups have come to expect. The whaling nations hate and ignore their attempts at setting quotas on the number of whales to be killed, while animal rights activists believe the IWC to be complicit in allowing the continuation of whaling when such creatures are on the brink of extinction. There is also the question of enforcement: it is impossible to force a country to abide by

its rulings. Really, only pressure from the public and other nations can force changes.

408.Johnsen, Arne O., and Tonnessen, J. N. *The History of Modern Whaling* (4 volumes in Norwegian, one volume in the English translation). Berkeley: University of California, 798 pp., 1982. (0520039734) Read a pro-whaling article at [www.naiaonline.org/newwhal.html]. Until the nineteenth-century, whaling did not have much of an impact on whale populations, with the exception of the slow, coast-hugging right whale, which was an easy target. The advent of steam ships, in conjunction with the new demand for oil led to a huge increase in whaling. Now, explosive harpoons and sonar location devices make whaling rather simple, and the whale populations have plummeted. One tactic has been the intentional harpooning of baby whales, whose parents or pod usually return to help, and then are speared themselves.

409.Jones, S. D. Morgan, ed. *Quality and Grading of Carcases of Meat Animals*. Boca Raton, FL: CRC, 234 pp., 1995. (0849350239) See animal by-product carcass rules at [www.maff.gov.uk/animalh/by-prods/appdx-c1.htm]. My understanding is that every part of an animal is used, even bones and skin and organs, which are not put into human food. Some goes to pet food, some to making clothing, and some becomes glue.

410.Kadletz, Edward. *Animal Sacrifice in Greek and Roman Religion*. Seattle: University of Washington, 1976. Note: this item should be in the chapter on "Animal Speculations," however, I wished to remind the readers that religions do also have fatal uses for animals, which will be more prominently noted in the "Animal Speculations" chapter. Kadletz is a professor of classics at Ball State University. Animal sacrifice has been found in most religions, not just Greek and Roman ones. Usually the theory behind this action is that the killing of an animal "appeases" the god or gods. In religions like Judaism, animals were actually a "stand-in" to pay for the sins of the human who offered it. Animal sacrifice has largely faded away due to the growth of Christianity (where Jesus' sacrifice ended the need for animal sacrifice) and the lessening of animistic religions.

411.Kay, William J., et al., eds. *Euthanasia of the Companion Animal: The Impact on Pet Owners, Veterinarians, and Society*. Philadelphia: Charles Press, 288 pp., 1998. (0914783246) A total of 37 contributors, including veterinarians, psychologists, pet owners, and others. Kay worked as chief of staff in the Animal Medical Center in New York City. See an Animal Welfare Information Center (AWIC) bibliography at [www.nal.usda.gov/awic/pubs/oldbib/srb9801.htm]. The theory behind euthanasia of a companion animal (a pet) is that the human owner is seeking to end the pain of the animal he or she cares about. Usually death is administered by lethal injection or poison gas, when an animal has become old or has been seriously injured. Pets become rather like members of the family, and so the death of a pet is usually a very emotional time for the family members. Do we have the right to make a life or death decision on behalf of an animal? Most pet owners believe that we do.

412. Kerasote, Ted. *Bloodties: Nature, Culture, and the Hunt.* New York: Kodansha International, 277 pp., 1993. (1568360274) The author says yes, we can kill animals for food, but no, not just to take trophies. He is not entirely pro or con on the hunting question. Read about the author at [www.denverpost.com/books/book172.html]. Read an article by the author at [www.colgate.edu/scene/jan1998/primal.html]. This is a very interesting book. What is the purpose of trophies? Is it the proud hunter displaying the dead animal (whole or part) for the admiration of friends, or simply a memento, like a souvenir picked up at a tourist shop? Taxidermy is a rather expensive procedure. I recently heard a traffic announcer on the radio observing a pair of men along a highway cutting the antlers off of a road-kill deer (with a hack-saw), presumably so that they could claim this trophy without even successfully hunting the game for themselves. How about the big Marlin mounted on grandfather's wall? Why display such things? Beauty? Reminders? Boasting?

413. Khodabandehloo, Koorash, ed. *Robotics in Meat, Fish and Poultry Processing.* London: Chapman & Hall, 1992. (0751400874) The author is an engineering professor in Great Britain. Read the table of contents at [www.engr.ucdavis.edu/~cfe/books/robotmfp.htm]. How might the rise of robotics affect animal welfare? If robots take over meat-processing lines, will the robots be able to recognize strange situations or alter procedures to avoid animal pain, if, for example, the animal is not properly placed on a production line?

414. Kneen, Brewster. *Farmageddon: Food and the Culture of Biotechnology.* Gabriola Island, Canada: New Society Publishers, 240 pp., 1999. (08657-13944) The author is a Canadian expert on the food system, and has written several articles and books about biotechnology. He writes about the ignorance of consumers about the vast amount of bioengineering that is going on in our food supply. "[W]hat's really happening down on the farm."

415. Kramer, David C. *Animals in the Classroom: Selection, Care, and Observations.* Reading, MA: Addison-Wesley, 234 pp., 1988. (02012-0679x) See an article about Massachusetts laws on animals in education at [mspca.org/ CLAW/SecondaryED/CLAW_txt_IA_SecondaryED.html] and

a resource page at [www.vetmed.ucdavis.edu/Animal_Alternatives/pre college.htm].

416. Krech, Shepard, III, ed. *Indians, Animals and the Fur Trade: A Critique of Keepers of the Game*. Athens: University of Georgia, 207 pp., 1981. (0820305634) Papers from a 1979 symposium. The author, an anthropology professor from Brown University, also wrote a 1984 book on the sub-arctic fur trade. See a site about early American fur trading at [www.xmission.com/~drudy/amm.html]. Sometimes animal rights supporters overgeneralize the "goodness" of groups who seem to support their positions. Native Americans are certainly not environmental villains, and they had many wise practices in regard to the use of wildlife resources. Nevertheless, some tribes of Native Americans were complicit in assisting the "white man" in decimating some animal populations, like buffalo and beavers. Any time there is money to be made (by collecting pelts, in this case), some people of all races would join in to share in the profits.

417. Krulisch, Lee, ed. *Implementation Strategies for Research Animal Well-Being: Institutional Compliance with Regulations*. Bethesda, MD: Scientists Center for Animal Welfare, 178 pp., 1992. Proceedings of a two-day conference on USDA regulations. The publisher's website (this author is the director) is at [www.scaw.com]. See a sample of such regulations at [www.macarthur.uws.edu.au/research/ethics/animal.html]. Without going out on a limb, I think it would be fair to say that any time a government agency creates volumes of regulations (right or wrong), it is difficult to sift through them and implement them.

418. Kuker-Reines, Brandon. *Environmental Experiments on Animals: A Critique of Animal Models of Hypoxemia, Heat Injury and Cold Injury*. Boston: New England Anti-Vivisection Society, 40 pp., 1984. The author says that our money is wasted on such experiments because we cannot extrapolate from animals to humans in such cases. Some articles by the author can be found at [www.geocities.com/CollegePark/8273/fb-index. htm]. Certainly these tests can tell us interesting things about how humans may react (and later heal) when they suffer serious heat or cold injuries. Still, these are among the more painful of animal experiments — broiling or freezing an animal to check its responses and abilities. How many such experiments are needed and how often? How many companies must do this?

419. Lafollette, Hugh, and Shanks, Niall. *Brute Science: Dilemmas of Animal Experimentation*. Philosophical Issues in Science series. New York: Routledge, 320 pp., 1997 (originally published in 1993). (0415131138) The authors say that vivisection is useful for biomedical research but not

for medical progress. A very interesting idea is that the more reliable the animal model, the less ethical it becomes to experiment on that animal. See several articles by Lafollette at [www.etsu.edu/philos/faculty/hugh.htm]. See a biography of Shanks at [www.etsu-tn.edu/philos/faculty/shanks.htm]. Is the human potential-good more important than the animal's certain-good? We may receive helpful information from a given experiment, but the animal(s) will certainly receive harm from that experiment. Some believe that humans always take precedence in any ethical question, while others say that humans do not naturally own the right of precedence over other species.

420.Langley, Gill, ed. *Animal Experimentation: The Consensus Changes.* New York: Chapman & Hall, 268 pp., 1990. (041202411x) A collection of essays outlining the change in morality we are undergoing in the world today on vivisection. Read an interesting article on the personality characteristics of animal activists and their opponents at [www.psyeta.org/sa/sa1.2/broida.html]. We really have seen a remarkable change in the late twentieth century. Though vivisection has by no means been abolished (or perhaps has even been reduced), just the fact that scientists will now admit that there is an ethical element to the activity is a major step. From this admission there can be discussion, and perhaps even consensus some day.

421.Langley, Gill. *Blinded by Science: The Facts about Sight Deprivation Experiments on Animals.* Tonbridge, England: Animal Aid Society, 1987. Read an article by the author at [www.newscientist.com/nsplus/insight/animalexperiments/langley.html]. This is another of the more brutal looking experiments on animals: the removal or intentional damaging of the eyes. Does the good information we may gain from the tests justify the harm done to the creatures?

422.Lansbury, Coral. *The Old Brown Dog: Women, Workers, & Vivisection in Edwardian England.* Madison: University of Wisconsin, 212 pp., 1985. (0788150928) In 1907 a number of diverse groups including anti-vivisectionists, suffragettes, and trade unions joined forces to fight London University medical students over vivisection on a dog. Sometimes we find allies in the strangest places. In this case, several groups with completely different interests and purposes came together to oppose this experiment. From time to time you may see this in Washington, DC — a coalition of odd "bedfellows" cooperating to promote or oppose some cause. Animal rights activists (and their opponents) would be wise to consider chatting with other organizations to see if they might work together in certain cases.

423.Laycock, George. *The Hunters & the Hunted: The Pursuit of Game in America from Indian Times to the Present*, reprint ed. Upland: DIANE

Publishing Co., 280 pp., 1990. (0788150928) See a list of modern hunting books that are available at [fordinfo.com/bookstore/ad.htm].

424.Lembeck, Fred. *Scientific Alternatives to Animal Experiments.* Ellis Horwood Series in Biochemistry and Biotechnology. New York: Chapman & Hall, 300 pp., 1990. (0412027712) Both sides are discussed, but all of the contributors seem to be in favor of vivisection. Find excellent resources for alternatives at [www.sph.jhu.edu/~altweb/science/pubs/altresources/newsletters.htm]. There are several types of nonanimal methods of testing chemicals now, including computer models and cell cultures. Some wonder, however, if these methods are as effective and informative.

425.Lyman, Howard F. *Mad Cowboy: Plain Truth from the Cattle Rancher Who Won't Eat Meat.* New York: Scribner, 192 pp., 1998. (0684845164) This guest on the Oprah Winfrey television show had the meat industry suing Oprah for defamation. This former rancher is funny and persuasive, saying that meat-eating may cause cancer, heart disease, obesity, and Mad Cow Disease. He tells of many questionable practices by ranchers that can pollute the meat. He recently became a vegetarian. Read an excerpt at [www.madcowboy.com/excerpt.htm]. Reviews at [160.79.243.32/issues/cc116/howardlyman.html], [www.montelis.com/satya/backissues.jun98/lyman.html], and [www.ivu.org/books/reviews/mad-cowboy.html]. The intensity of the reaction by the cattle-industry to Lyman's comments was remarkable. Part of the reason is the fear of Mad Cow Disease, which crippled Great Britain's beef industry for several years. So far there has been little or no evidence of the affliction in America, at least from beef. A similar disease has plagued a few herds of sheep in the United States, but thus far it does not seem to have affected any humans.

426.MacDonald, Mclody. *Caught in the Act: The Feldberg Investigation.* Northvale, NJ: Jon Carpenter, 1994. (189776605x) See more information about the infamous Feldman case (in Britain) at [www.nsplus.com/nsplus/insight/animal/trenches.html]. See also a good review of the book on the Amazon.com bookstore website.

427.Mahoney, James. *Saving Molly: A Research Veterinarian's Hard Choices for the Love of Animals.* Chapel Hill, NC: Algonquin Books, 252 pp., 1998. (1565121732) After a long effort to save a pet, the author wrestles with the ethics of the experiments he performs on other animals but decides they are needed. Reviewed at [www.workmanweb.com/algonquin/a7_7.html].

428.Mahoney, Ruth Carmela. *Animal Testing Alternatives in Health Sciences: Index of New Information with Authors, Subjects & Bibliography,* revised ed. Washington, DC: ABBE, 147 pp., 1994. (0788303686) Related links at [www.frame-uk.demon.co.uk/Links/3rslinks.htm].

429. Martin, Ann N., and Fox, Michael W. *Food Pets Die For: Shocking Facts about Pet Food.* Troutdale: New Sage Press, 148 pp., 1997. (0939165317) A seven-year study showing that much pet food is waste unfit for human consumption, made up of leftovers from slaughterhouses. This bad meat presents dangers to pet health. The industry is entirely unregulated but worth $20 billion per year. Includes an interesting chapter on the dangers of Mad Cow Disease to pets. See [www.teleport.com/~newsage/books/petfood.html] and [home.earthlink.net/~astrology/petfood.html]. The oft-told stories about the sort of leftover meat products that go into hot dogs apparently are nothing compared to the pet food contents. This is particularly nasty in domestic animal feed, which is often just animal skin and brain and tail ground up into protein powder. The rising price of grain led farmers to look for other methods of getting their animals protein. Unfortunately this is also the way that cows got Mad Cow Disease in Britain, by eating scraps of brain from other cows.

430. Mason, Jim, and Singer, Peter. *Animal Factories: The Mass Production of Animals for Food and How it Affects the Lives of Consumers, Farmers and Animals*, revised ed. New York: Crown, 174 pp., 1990 (originally published in 1980). (051753844x) A remarkable and persuasive book showing the sad conditions of modern factory farms. Read an article by Mason at [www.montelis.com/satya/backissues/jun97/animals.html]. This is the book that started my thinking about the ethics of food animal production. Before, I had no idea what things were like. My grandfather had a farm with beef cows that roamed in big pasture fields; I just assumed that was how all cows were raised. Now such family farms are fairly rare; big industry owns most of the modern farms.

431. McCoy, J. J. *Animals in Research: Issues and Conflicts.* An Impact Book. New York: Franklin Watts, 128 pp., 1993. (0531130231) Written for young adults. Some reviewers say that it is biased on the pro vivisection side since it never mentions ideas of animal suffering. Read a brief article by a "moderate" on the subject of vivisection at [www.paweekly.com/PAW/morgue/cover/1996_Nov_6.ANIMAL2.html].

432. McKenna, Carol. *Fashion Victims: An Inquiry into the Welfare of Animals on Fur Farms.* World Society for the Protection of Animals (WSPA), 1998. The last chapter is written by Andrew Linzey. Read the book online, starting at [www.infurmation.com/investig/fash_vic/wspafa01.htm].

433.McNally, Robert. *So Remorseless a Havoc: Of Dolphins, Whales and Men.* Boston: Little, Brown, 268 pp., 1981. (0316562920) Topics include cetacean intelligence and whaling history. See a good related site at [www.dolphin-institute.com]. Whales are the largest living creatures on Earth. Scientists now says that whales are also among the most intelligent animals. One of the newer issues about whales is about disturbing their habitat. The U.S. Navy has designed some new super-sonar that can detect submarines at great distances, but critics say that it is on the same frequencies that whales use to communicate, and may "hurt their ears" or disrupt their communications.

434.Medical Research Modernization Committee. *A Critical Look at Animal Research,* 5th ed. New York: Medical Research Modernization Committee, 17 pp., 1998. Read the booklet at [www.mrmcmed.org/critcv.html].

435.Medical Research Modernization Committee. *Shortcomings of AIDS-Related Animal Experimentation.* New York: Medical Research Modernization Committee, 1996. Read the work at [111.mrmcmed.org/aids.html]. Traditionally, "liberal" organizations that support animal rights would also support funding of AIDS research, but in this case, the camps are in opposition. People with AIDS probably do hope that animal research will lead to a cure, but this is incompatible with the standard anti vivisectionism of most animal rights activists.

436.Mench, Joy A., and Krulisch, Lee, eds. *The Well-Being of Agricultural Animals in Biomedical and Agricultural Research.* Bethesda, MD: Scientist Center for Animal Welfare, 112 pp., 1992. Proceedings of a conference in 1990. These editors have compiled a number of books on animal well-being, including a 1989 book regarding primates. Read an article by Mench at [forum.ra.utk.edu/mench.htm]. See an article about creating comfortable quarters for animals at [www.animalwelfare.com/pubs/cq/cq_page.html].

437.Mettler, John J., Jr. *Basic Butchering of Livestock & Game.* Pownal: Storey, 208 pp., 1989. (0882663917) The book has more than a hundred illustrations. See an interesting article about butchers at [metroactive.com/papers/metro/11.02.95/meat-9544.html]. Some methods of killing an animal for food are more painful than others. Kosher butchering is controversial, since it requires the animal be bled to death by a throat slash; but some say this is painless.

438.Metz, J. H. M., and Groenestein, C. M. *New Trends in Veal Calf Production.* EAPP Publication # 52. EAPP, 154 pp., 1991. Read a good article at [www.vetmed.ucdavis.edu/vetext/INF-AN_Veal95France. html]. The lives of veal calves are among the most restricted of the factory farm animals. In order to keep their meat white and soft, they are kept in small pens, unable to walk or turn around, and fed a strict diet to keep them anemic. They are slaughtered after several weeks. Most veal calves are males from milk farms; since male cows do not produce milk, they are useless to the milk farmer.

439.Michigan Society for Medical Research. *Animal Research & Testing* (video). Ann Arbor: Michigan Society for Medical Research, 1992. This organization has also produced other relevant videos, including *Animal By-Product* (28 minutes, 1994) and *Man and Animals: Partners for Life* (23 minutes, 1990). *Animal Research & Testing* is 29 minutes in length. See the publisher's website at [www.med.umich.edu/mismr].

440.Monamy, Vaughan. *Animal Experimentation: A Student Guide to Balancing the Issues*. Glen Osmond, Australia: ANZCCART, 1996. (09586-82100) See the table of contents at [www.adelaide.edu.au/ANZCCART/AnimalExperimentation.html].

441.Moretti, Laura A., ed. *All Heaven in a Rage: Essays on the Eating of Animals*. Chico, CA: MBK Publishing, 80 pp., 1999. (1884873146) Moretti is the founder and editor of *The Animals' Voice* magazine. Well-chosen and concise essays, including one by Paul Ehrlich. Read an article by the author at [factoryfarming.nettinker.com/farmsanctuary/newsletter/news_moretti3.htm].

442.Mowat, Farley. *A Whale for the Killing*, reprint ed. Mattituck, NY: Amereon Ltd., 213 pp., 1991 (originally published in 1972). (0891908226) The sad story of an 80-ton fin whale that became trapped in a Newfoundland lagoon and was slowly shot and cut to death by uncaring locals. This was made into a television movie. Read a good biographical article on Mowat at [www.salonmagazine.com/people/bc/1999/05/11/mowat/index.html].

443.Mowat, Farley, and Brooks, Cristen. *Sea of Slaughter,* reprint ed. New York: Chapters Publishing Ltd., 438 pp., 1996 (originally published in 1984). (1576300196) This famous Canadian author says that *Sea of Slaughter* was his most important work. Shows the current danger of extinction to Arctic Circle birds, seals, fish, and whales. Lists many of the creatures that are already gone. Read more about the author at [www.tceplus.com/mowat.htm]. The animals of the Arctic seem to be more vulnerable than others to human invasion, perhaps because they are easier to spot on the ice or because there are fewer spots where they may bear their young. The Great Auk, a giant penguin, was made extinct in the nineteenth century because people stole their giant eggs off the beaches where they lay. What better way to exterminate a species than to steal and eat all of its young? There was simply no new generation to continue to breed once all the eggs had been collected.

444. National Research Council. *Guide for the Care and Use of Laboratory Animals*, 7th ed. Washington, DC: National Academy Press, 125 pp., 1996 (originally published in 1963). (0309053773) First issued in 1963 and revised every few years since then. Intended to show standards for "well-being" of experimental research animals. Includes sample policies, veterinary care standards, and architectural designs. Read the whole book at [www.nap.edu/catalog/5140.html].

445. National Research Council. *Recognition and Alleviation of Pain and Distress in Laboratory Animals.* Washington, DC: National Academy Press, 160 pp., 1992. (0309042755) Guidelines for researchers on how to reduce discomfort in lab animals. Read the table of contents at [books.nap.edu/catalog/1542.html]. Unlike past decades, most scientists now acknowledge that animals do feel pain and that pain should be reduced when possible. This is actually a big step for the conservative scientific community, which has clung rigidly to the ideas of Rene Descartes' "beast-machine" theory: that animals are incapable of feeling pain.

446. National Research Council. *The Use of Drugs in Food Animals: Benefits and Risks.* Washington, DC: National Academy Press, 276 pp., 1999. (085-1993710) Why and how drugs are used in food animals; and what human pathogens may arise from such drugs. How microbes build resistance by mutation to some drugs. Read the whole book at [books.nap.edu/html/foodanim/]. This has become a major issue in recent years, as the European Union is refusing to buy American beef due to our extensive use of drugs in cattle. The Food and Drug Administration assures us that the drugs are safe. Many of the drugs used in cattle are a direct result of overcrowded conditions in factory farms, which require heavy doses of antibiotics to keep any diseases from spreading rapidly through the huge herds. From time to time, when a dangerous infection does take hold in a herd, the whole herd must be destroyed (as seen in Great Britain in the Mad Cow Disease affair).

447. Nilsson, Greta, et al., eds. *Facts about Furs.* Washington, DC: Animal Welfare Institute, 257 pp., 1988. History of fur, types of traps. Read the foreword and table of contents at [www.animalwelfare.com/pubs/bk-conts.htm].

448. Nyberg, Cheryl Rae, Porta, Maria A., and Boast, Carol. *Laboratory Animal Welfare: A Guide to Reference Tools: Legal Materials, Organizations, Federal Agencies.* Twin Falls, MN: Boast Nyberg Books, 391 pp., 1994. (0961629398) See a good site on animal care in laboratories at [www.iacuc.org]. Links at [www.faseb.org/aps/classroom.htm].

449. Orlans, F. Barbara. *In the Name of Science: Issues in Responsible Animal Experimentation*, reprint ed. New York: Oxford University, 297 pp., 1996 (originally published in 1993). (019510871x) The author is a vivisectionist (research fellow at Georgetown University) who says there must be some limits on experimental use of animals. Find a bibliography for lab animal workers at [www.animalwelfare.com/Lab_animals/biblio/iacuc2.htm]. Now that some scientists have admitted that there is an ethical dimension to be respected with regards to animals, a few have themselves called for reductions in the number and type of vivisections to be performed.

450. Ortega y Gasset, Jose. *Meditations on Hunting: Provocative Insights into Anthropology & Ecology by the Great Spanish Thinker*, reprint ed. Bozeman, MT: Wilderness Adventure Sports Press, 152 pp., 1996 (originally published in 1942). (1885106181) The publisher says that this is "the most quoted book in sporting literature." A philosophical defense of hunting by a Spaniard. Read one chapter and the table of contents at [www.wildadv.com/classic.html]. See a related article at [www.ool.com/wildlife-forever/symposium/fritzell.html]. The selections seem to be very metaphysical and animistic in their glorification of the primal hunt. It sounds like nationalistic fervor in the defense of a country; as if man is accomplishing a great mission and protecting the world by partaking in sport hunting.

451. Overell, Bette. *Animal Research Takes Lives: Humans and Animals BOTH Suffer*. Wellington: New Zealand Anti-Vivisection Society, 368 pp., 1993. (0473018462) A refutation of a pro vivisection booklet printed by the New Zealand government in 1990. Read an extensive description at [www.nzavs.org.nz/inARTL.html].

452. Page, Tony. *Vivisection Unveiled: An Expose of the Medical Futility of Animal Experimentation*. Oxford: Jon Carpenter, 160 pp., 1998. (1897766319) The author founded the UK Anti-Vivisection Information Service. Reviewed at [www.werple.net.au/~antiviv/review.htm#VivisectionUnveiled].

453. Partners in Research. *Biomedical Research: Is It Really Necessary?* (video). Ottawa: Partners in Research, 1992. A 32-minute Canadian video

made for high schools, in favor of vivisection. See the publisher website at [www.pirweb.org/faq.html].

454.Paton, William D. M. *Man and Mouse: Animals in Medical Research*, 2nd ed. New York: Oxford University, 288 pp., 1993 (originally published in 1984). (0192861468) The author believes that animals have no rights because they do not participate in society. One interesting justification for cosmetic testing is that cosmetics are not trivial because they may "conceal" unsightly birthmarks (and thus promote good mental health). See an article in favor of this work at [www.nsplus.com/nsplus/insight/animal/barricade.html]. Some distinction should be made about the views of those who claim that animals do not have any rights. This view does not necessarily mean that they believe it is morally acceptable to be cruel to animals. Their views simply do not grant the same kind of rights as we ascribe to humans also to the animals. Some Christians, for instance, believe that "rights" are a bogus idea all around, that no one has rights except God, and any kindnesses we show to humans and animals are simply obedience to God's commands. So the "rights" issue is not the same as the treatment that the authors may wish for the animals.

455.Peden, James A. *Vegetarian Cats and Dogs*. Troy, MT: Harbingers of a New Age, 238 pp., 1998. (0941319016) See a page for this book at [www.montanasat.net/vegepet/bookvcd.html], and an article at [www.earthsave.bc.ca/articles/articles/ethics/vegetarian_cats_and_dogs.html]. A few vegans and vegetarians have begun to insist that even their pets be part of their own meatless diets. It is rather controversial though, and difficult. Pets cannot understand the morality of a meatless diet and are prone to desire meat. And unlike humans, their teeth and digestive systems are those of predators, with sharp canine teeth. Promoters say that the animals can thrive on a vegetarian diet.

456.Penman, Danny. *The Price of Meat: Salmonella, Listeria, Mad Cows — What Next?* North Pomfret, VT: Victor Gollancz, 240 pp., 1997. (05750-63440) Briefly reviewed at [www.animalwelfare.com/farm/bkrviews.htm].

457.Phillips, Mary T., and Sechzer, Jeri A. *Animal Research & Ethical Conflict: An Analysis of the Scientific Literature, 1966-1986*. New York: Springer Verlag, 250 pp., 1989. (0387963957) The authors present an analysis of responses of research scientists to the experimentation controversy and a 300-item bibliography. Read an article by Phillips at [www.psyeta.org/sa/sa1.1/phillips.html]. See an article with both pro and con views at [www.psyeta.org/sa/sa3.1/paul.html].

458. Poole, T. B., ed. *The UFAW Handbook on the Care and Management of Laboratory Animals*, 7th ed. Oxford: Blackwell, 1999. Two volumes by the Universities Federation for Animal Welfare group. Find a bibliography at [www.lib.umd.edu/UMCP/MCK/GUIDES/animal_welfare.html].

459. Porphyry. *On Abstinence from Animal Food.* Ancient Commentaries on Aristotle series. Ithaca, NY: Cornell University, 196 pp., 1999. (08014-36923) Religious animal sacrifice does not justify meat- eating, says this author. Many animals are rational and justice is owed to rational creatures. See a brief biography and some quotes at [ivu.org/people/history/porphyry.html]. This is one of the oldest discussions of animal rights issues to be found, the third century A.D.!

460. Posewitz, Jim. *Beyond Fair Chase: The Ethic and Tradition of Hunting.* Helena, MT: Falcon Publishing Co., 118 pp., 1994. (1560442832) The ethics of hunting and shooting. This book is used in many hunter education courses as a text. Read an excerpt at [www.hcn.org/1995/dec11/dir/Opinion_For_this_h.html].

461. Pratt, Dallas. *Alternatives to Pain in Experiments on Animals.* New York: Argus Archives, 283 pp., 1980. See the whole book online at [www.animalwelfare.com/lab_animals/pratt/prat-idx.htm].

462. Recreational Fishing Alliance. *Gladiators of the Deep: Rape of a Resource* (video). Washington, DC: Recreational Fishing Alliance, 1998. A nicely prepared video (about 20 minutes) showing how driftnetting is decimating the world's oceans of fish and marine mammals. Also shows the cruel practice of shark definning: fins are a delicacy in Asia. The Recreational Fishing Alliance is a group of sport fishermen who want to conserve salt-water species for future enjoyment. Their homepage is at [www.tbs-tournaments.com/RFA.htm]. Read an article supporting the tuna industry at [www.naiaonline.org/dolphin2.html].

463. Reilly, J. S., ed. *Euthanasia of Animals Used for Scientific Purposes.* London: Universities Federation for Animal Welfare, 71 pp., 1993. (064-611803x) Methods of euthanasia and training advice for involved persons. See a content summary at [www.adelaide.au.edu/ANZCCART/Euthanasia#anchor1168799]. Not all animal tests are fatal, some merely inflict injuries (which may heal), and some merely observe behavioral reactions. So what does a lab do when the tests are finished? The common practice has been to kill the unneeded animals. However, some folks have created animal parks to care for some of these animals, especially the monkeys. This is similar to some of the parks created for older homeless circus animals, like elephants.

464. Reinhardt, Christoph A. *Alternatives to Animal Testing: New Ways in the Biomedical Sciences*. New York: Wiley, 182 pp., 1994. (352730043) How the "Three R's" of refinement, reduction and replacement are working to improve the work of vivisection. See an article on the third R, refinement, at [www.vetmed.ucdavis.edu/Animal_Alternatives/REFINE.htm]. Read a list of alternatives that are not well known by vivisectionists at [www.nsplus.com/nsplus/insight/animal/experiment.html].

465. Reitman, Judith. *Stolen for Profit: The True Story Behind the Disappearance of America's Beloved Pets*, revised ed. New York: Kensington, 307 pp., 1995 (originally published in 1992). (082174951x) The author says that many missing pets are not hit by cars but stolen by people who then sell them to vivisection laboratories for testing. Apparently there is money to be made this way. Some of the people who answer newspaper ads for kittens and puppies "free to good homes" are actually taking them to labs. Critics say that this is simply an "urban legend" which has no real basis in fact: like the early 20th century scare about young ladies being lured into white slavery. Read a review at [www.api4animals.org/Publications/AnimalIssues/1995-Summer/Reviews95-02.htm] and a list of dangers for pet owners at [www.petaonline.org/cmp/crcafs7.html].

466. Rifkin, Jeremy. *Beyond Beef: The Rise and Fall of the Cattle Culture*. New York: Dutton, 353 pp., 1993. (0452269520) An indictment of the cattle culture that has "warped our world." He says that the transition of arable land from production of human food to production of cattle feed leads to starvation, poverty, global warming, and disease. See an article by the author about genetic engineering at [www.emagazine.com/may-june_1998/0598feat2.html]. Reviewed at [www.psyeta.org/sa/sa1.2/birke.html]. Read an article on this topic at [www.enviroweb.org/mcspotlight-na/media/reports/beyond.html]. A huge amount of the grain produced by American farms is not consumed by humans but by the cattle whom the humans later eat. Purportedly, the meat produced has only one-tenth the nutritional value of the grain, had it been used by humans. This, if true, would make meat quite a luxury and not the staple food we consider it to be today.

467. Robbins, John. *Diet for a New America*, 2nd ed. Washington, DC: People for the Ethical Treatment of Animals, 448 pp., 1998. (0915811812) Why a vegetarian diet can save individual people and the whole planet. Nominated for a Pulitzer Prize. Read Chapter Ten online at [arrs.envirolink.org/ar-voices/iron.html]. Read an interview at [www.healthy.net/library/interviews/redwood/jrobbins.htm]. Find a good list of the "facts" from the Robbins book at [www2.utep.edu/~best/handout.htm].

468. Rollin, Bernard E. *The Frankenstein Syndrome: Ethical and Social Issues in the Genetic Engineering of Animals*. Cambridge Studies in Philosophy and Public Policy series. Cambridge: Cambridge University, 241 pp., 1995. (0521478073) A very interesting analysis of genetic engineering, its problems and its benefits. The opening chapter is also a very thorough critique of the modern scientist's attitude which says that science need not consider any ethical questions during its work. Reviewed at [pages.inrete.it/immunoblack/1/frank.html]. See a bibliography of genetic engineering resources at [www.a-ten.com/alz/ge.htm]. Genetic engineering of humans and animals must be one of the most difficult issues we will face over the next centuries. There is no way of predicting accurately where it will lead. Rollin rightly points out that there are both very good and very bad possibilities. We must try to keep the technology and practice going in the directions that we consider to be good. Certainly, this will create major changes in our lives. We may live longer, or be more healthy, or pick and choose the characteristics we want for our children.

469. Rollin, Bernard E., and Kesel, M. Lynne, eds. *The Experimental Animal in Biomedical Research: Care, Husbandry, and Well-Being: an Overview by Species*, 2 volumes. Survey of Scientific and Ethical Issues for Investigators series. Boca Raton, FL: CRC Press, 1995. (0849349818 and 0849349826) See an article about all types of animals in research at [netvet.wustl.edu/species/].

470. Rowan, Andrew N. *Of Mice, Models, and Men: A Critical Evaluation of Animal Research*. Albany: State University of New York, 323 pp., 1984. (0873957768) This biochemist talks about vivisection, and concludes that we should greatly reduce the number of experiments, with the ultimate with goal of abolishing it. "Where the scientist, animal welfare advocate, and abolitionist differ is over the feasible time scale for the elimination of animal research, and the vigour with which the goal should be pursued." Read an article by the author at [altweb.jhsph.edu/science/meetings/pain/rowan.htm]. This would be another moderate proposal: we will gradually lessen the animal testing, and in the end abandon it — not an immediate all-or-nothing demand.

471. Ruesch, Hans. *Naked Empress, or The Great Medical Fraud*, 3rd ed. Klosters, Switzerland: CIVIS, 260 pp., 1992 (originally published in 1982). (3905280027) We should improve health by preventing disease, abolishing vivisection, and reducing pharmaceutical drugs. How the media is complicit in spreading pro vivisectionist propaganda. Reviewed at [www.geocities.com/CollegePark/8273/empress.htm]. Read an excerpt at [www.pnc.com.au/~cafmr/reviews1.html#naked].

472. Ruesch, Hans. *1000 Doctors (and Many More) Against Vivisection.* Massagno, Switzerland: CIVIS, 290 pp., 1989. Lots of quotations in chronological order. Much of the book can be read at [www.geocities.com/CollegePark/8273/hr-index.htm]. See excerpts at [www.pnc.com.au/~cafmr/online/ research/dav.html].

473. Ruesch, Hans. *Slaughter of the Innocent.* Swain, NY: Civitas, 446 pp., 1991 (originally published in 1976). (0553111515) A critique and history of vivisection. The author says it does more harm to humans than good. Very controversial: the major publisher, Bantam, dropped it. The conclusion contains a very thinly veiled threat to violence against vivisectors. Read the foreword at [www.geocities.com/CollegePark/8273/mendel.htm]. What should we make of threats of violence? Certainly people have strong feelings on these issues. Let's say for the sake of discussion, that animals have the same right to life that people do. Are we then justified in threatening to physically harm those who harm animals, or actually doing violence (not just threatening it)? So far, all the direct actions sanctioned by animal rights groups have been violent only against property, not against people.

474. Rupke, Nicolaas A., ed. *Vivisection in Historical Perspective.* Wellcome Institute Series in the History of Medicine. New York: Routledge, Kegan & Paul, 373 pp., 1990 (originally published in 1987). (0709942362) Fifteen documented essays on the controversy. The author is a history professor at Vanderbilt University. See related articles at [funnelweb.utcc.utk.edu/~ilsmith/SVMEProc97.html].

475. Ryder, Richard D. *Victims of Science: The Use of Animals in Research,* 2nd ed. London: National Anti-Vivisection Society, 279 pp., 1983 (originally published in 1975). (0905225066) This book may have coined the term "speciesism" (one source that claims Peter Singer uses the term in his 1975 work *Animal Liberation*). The author (former chairman of the RSPCA) says that we can only justify vivisection in benefits to the same individual, not in an ephereal or supposed benefit to a whole species or to all mankind. Read an article by Ryder (reviewing another book) at [www.psyeta.org/sa/sa3.2/ryder.html].

476. Salem, Harry, and Katz, Sidney A., eds. *Advances in Animal Alternatives in Safety and Efficacy Testing,* revised ed. London: Taylor & Francis, 456 pp., 1997 (originally published in 1995). (1560326239) More than 40 contributing scientists. Read a publisher review and table of contents at [www.taylorandfrancis.com/BOOKS/toxicology/advaniml.htm].

477. Sanderson, Ivan T. *A History of Whaling*, reprint ed. New York: Barnes & Noble, 423 pp., 1993 (originally published in 1953). This book was previously titled *Follow the Whale*. Briefly reviewed at [www.physics.helsinki.fi/whale/intersp/pages/book2.htm]. See a pro whaling site at [pw2.netcom.com/~pokey647/main.html].

478. Savory, C. J., and Hughes, B. O., eds. *Proceedings of the Fourth European Symposium on Poultry Welfare*. London: Universities Federation for Animal Welfare, 318 pp., 1993. (0900767839) Forty papers. See an article about turkey production at [www.earthsave.bc.ca/articles/ethics/turkey.html].

479. Schar-Manzoli, Milly, and Keller, Max. *Holocaust: Vivisection Today Performed in 60 Laboratories, 150 Documented Accounts of Vivisection Crimes, Illegal Traffic, Clinical Breeding and Lobbies*. Casa Orizzanti, Switzerland: Swiss League for the Abolition of Vivisection, 1994. Read an article containing some information about Swiss vivisection training at [www.geocities.com/CollegePark/8273/anderegg.htm]. See a related article at [home.mira.net/~antiviv/issue495.htm].

480. Schell, Orville. *Modern Meat: Antibiotics, Hormones and the Pharmaceutical Farm*. New York: Random House, 337 pp., 1984. (039451890x) Potential risks with benefits on the new pharmaceutical farms. See an HSUS article about antibiotics at [www.ocpa.net/frames_library.htm]. Because of the densely populated factory farms, disease can spread quickly. To counter this possibility, the animals are given high doses of antibiotics. Hormones are injected or added to food in order to increase milk production or to help fatten up animals.

481. Schwartz, Marvin. *Tyson: From Farm to Market*. University of Arkansas Press Series in Business History, volume 2. Fayetteville: University of Arkansas, 170 pp., 1991. (1557281890) "[E]ntertaining and enlightening tribute to the Tyson vision and success." See a related article about the future of poultry birds at [www.upc-online.org/genetic.html]. Look at the Tyson website at [www.tyson.com]. Tyson is probably the largest distributor of poultry meat in the United States. Life for chickens has greatly changed for the worse in recent decades. Chickens routinely have their beaks and claws cut off to prevent fighting with other birds under the stress of the overcrowded environment. Male chicks are usually ground up into feed at birth, since females tend to become fatter broilers, and males cannot be laying hens for the egg industry.

482. Semencic, Carl. *The World of Fighting Dogs*. Neptune City, NJ: TFH, 287 pp., 1984. (0876665660) Read articles at [petnet.detnews.com/news/

fighting.htm] and [www.aspca.org/calendar/watfeaf7.htm]. This topic, along with cock-fighting and bear-baiting, is rarely addressed because it is an illegal sport (and lesser known). The basic idea is similar to the ancient Roman arena games: put two dogs in a cage, gamble on the event, and the winner is the dog still alive at the end. In earlier centuries these games were a joy of the rich, but it seems to have become more of a spectator sport for the poor and immigrants from Mexico in modern times. The events are held secretly in barns, generally in the southwestern U.S.

483. Sharpe, Robert. *The Cruel Deception: The Use of Animals in Medical Research.* Wellingsborough, England: Thorsons, 288 pp., 1989. A research chemist shows how vivisection usually is irrelevant to real advances in health. He says that most diseases have been cured or prevented not by vivisection but by improvements in sanitation and such things. Briefly reviewed at [home.mira.net/~antiviv/review.htm]. Read a related article on cancer research "fraud" at [www.healthwealthsolutions.com/health_news_and_information.htm#CancerResearch].

484. Sharpe, Robert. *Science on Trial: The Human Cost of Animal Experiments.* Sheffield, England: Awareness Books, 156 pp., 1994. The author is Director of the International Association Against Painful Experiments on Animals. Why does vivisection continue despite its many problems and much opposition? The introduction was written by activist and actor Sir John Gielgud. See an article by the author at [home.mira.net/~antiviv/article1.htm]. Read a review at [www.pnc.com.au/~cafmr/reviews1.html#science].

485. Sinclair, Upton. *The Jungle*, reprint ed. Reading, MA: Addison-Wesley, 353 pp., 1997 (originally published in 1907). (0321026020) On the hardships of immigrants (largely working in a meat packing plant, in this novel) around the turn of the twentieth century. Sinclair was accused of writing socialistic propaganda. This book was the first popular (and semi-fictional) expose of what goes on in meat packing plants. Inclusion of diseased or spoiled meats, along with other horrors, sent shockwaves across the nation and led to some investigations of the industry in the early twentieth century. Sinclair had sources inside the Chicago meat plants.

486. Skaggs, Jimmy M. *Prime Cut: Livestock Raising and Meatpacking in the United States, 1607-1983*, reprint ed. Upland, CA: DIANE Publishing Co., 280 pp., 1999 (originally published in 1986). (0788162748) See some pro beef articles at [www.beef.org/librref/rc_eb.htm].

487. Smith, Cynthia Petre, et al., eds. *Environmental Enrichment Information Resources for Laboratory Animals: 1965-1995.* AWIC Resource series,

No. 2. Washington, DC: Animal Welfare Information Center, 1995. (0900-76791x) Read the whole book at [www.nal.usda.gov/awic/pubs/enrich/intro.htm].

488. Smith, Jane A., and Boyd, Kenneth M., eds. *Lives in the Balance: The Ethics of Using Animals in Biomedical Research: The Report of a Working Party of the Institute of Medical Ethics.* New York: Oxford University, 352 pp., 1991. (0198547447) The 18 participants of this working group concluded that vivisection is a "necessary evil." See a pro vivisection article at [www.nsplus.com/nsplus/insight/animal/neglected.html].

489. Stange, Mary Zeiss. *Woman the Hunter.* Naples, FL: Beacon, 240 pp., 1997. (0807046388) Challenges the idea that men are hunters and women are gatherers. The author says that many women hunt. She also attacks ecofeminism, which equates hunting with rape. "All humans are dipped in blood." Read excerpts at [www.beacon.org/Beacon/authors/stange/stangex.html]. Reviewed at [www.montelis.com/satya/backissues/nov97/diana.html].

490. Stull, Donald D., et al. *Any Way You Cut It: Meat Processing and Small-Town America.* Rural America series. Lawrence: University Press of Kansas, 296 pp., 1995. (0700607218) Problems with agribusiness (often synonymous with factory farming) in small towns. Reviewed at [www.montelis.com/satya/backissues/sept96/bookreview.html].

491. Suarez, Jose, and Marks, John. *Life & Death of the Fighting Bull.* New York: Putnam, 127 pp., 1968. Photographic essay of a bull's life from birth to death: rearing, training, and final day of a bull's life in the arena. Read an article about bullfighting at [www.montelis.com/satya/backissues/nov96/interview.html].

492. SUPRESS. *Lethal Medicine* (video). Pasadena, CA: Students United Protesting Research on Sentient Subjects (SUPRESS), 1997. An expose on animal research, but includes interviews with scientists. Fifty minutes long, and the replacement for their 1987 film *Hidden Crimes.* See videoclips and transcripts at [www.animalresearch.org/TV.htm]. See the organization's homepage at [home.earthlink.net/~supress/].

493. Swan, James A. *In Defense of Hunting.* San Francisco: Harper, 304 pp., 1995. (0062512374) This work is a "love it or hate it" work, depending on your position. Swan calls animal rights activists a "new subspecies of human." See the author's page at [microweb.com/swan/James/James.htm]. See many quotes from the book at [www.acs.ucalgary.ca/~powlesla/personal/hunting/rights/ARtypes.txt]. Read an article by the author at

[www.intellectualcapital.com/issues/Issue166/item2019.asp]. Briefly reviewed at [www.wholeliving.com/redwood/envsocpol/defense1.html].

494. Thelestam, Monica, and Gunnarson, Anders, eds. *The Ethics of Animal Experimentation*. Proceedings of the 2nd CFN Symposium. Oxford: Blackwell, 269 pp., 1986. The authors recommend that scientists should act as stewards, but that experimentation is good and right. See an example of an institution's ethics policy at [www.research.ryerson.ca/~ors/policies/ethics-a.html].

495. Tuffery, A. A., ed. *Laboratory Animals: An Introduction for New Experimenters*, 2nd ed. New York: J. Wiley, 406 pp., 1995 (originally published in 1987). (0471952575) Feeding, housing, use of anesthesia, and new alternative procedures are discussed. Reviewed at [am.appstate.edu/top/dept/biology/arb/BookReviews.htm].

496. Universities Federation for Animal Welfare. *Euthanasia of Unwanted, Injured or Diseased Animals or for Educational or Scientific Purposes*. London: Universities Federated for Animal Welfare, 1986. (0785537589) See some statistics on euthanasia at [www.naiaonline.org/euthrisk.html].

497. Unknown. *Animal Research Facility Protection*. Washington, DC: Government Printing Office, 1991. See an article about the related law of 1991, the Animal Research Facility Protection Act S.544, at [www.nal.usda.gov/awic/newsletters/v3n3.htm].

498. VanZutphen, L. F. M., et al. *Principles of Laboratory Animal Science: A Contribution to the Humane Use and Care of Animals and to the Quality of Experimental Results*. New York: Elsevier, 389 pp., 1993. (04448-14876) See an article about the ethical treatment of lab animals at [www.apa.org/science/anguide.html].

499. VanZutphen, L. F. M., and Balls, M. *Animal Alternatives, Welfare and Ethics*. Developments in Animal and Veterinary Science, No. 27. New York: Elsevier, 1260 pp., 1997. (0444824243) Contains 160 papers presented at the World Congress on Alternatives conference in the Netherlands in 1996. Read an article about international views at [www.psyeta.org/sa/sa2.2/pifer.html].

500. Vyvyan, John. *The Dark Face of Science*. Marblehead, MA: Micah Publishing, 200 pp., 1988. (0916288226) The author is an archaeologist and Shakespearian scholar. He attacks animal experimentation using history, philosophy, and science. This book is a sequel to *In Pity and Anger*, giving the history of vivisection in the twentieth century up to the

1960s. He says that vivisection has led to a modern world full of cruelty. Read an article about secrecy in laboratories at [www.nsplus.com/nsplus/insight/animal/secret.html].

501. Vyvyan, John. *In Pity and In Anger: A Study of the Use of Animals in Science*. Marblehead, MA: Micah Publishing, 180 pp., 1988 (originally published in 1969). (0916288218) History of vivisection and anti-vivisection in the nineteenth century. The author says that "pity is a universal law, like justice," and thus science should not use cruel methods in achieving its objectives.

502. Warren, Louis S. *Hunter's Game*. Yale Historical Publication series. New Haven: Yale University, 227 pp., 1997. (0300062060) Poachers and conservationists in twentieth century America. The author says that at heart this is a struggle between rural residents and urban conservationists. The rural people see the urban voters as trying to turn wildlife into "public goods." Read a related article at [www.psyeta.org/sa/sa1.2/dahles.html].

503. Welsh, Heidi J. *Animal Testing and Consumer Products*. Washington, DC: Investor Responsibility Research Center, 176 pp., 1988. (0931035392) See a related site about cosmetics testing at [www.ddal.org/CCIC/], and the organization "Beauty Without Cruelty" at [animals.co.za/Orgs/BWC]. Find the publisher site at [www.irrc.org].

504. Whisker, James B. *The Right to Hunt*, revised ed. Bellevue, WA: Merril Press, 224 pp., 1999 (originally published in 1981). (0936783206) A professor from West Virginia University writes on historical, religious, legal, ethical, and practical aspects of hunting. Read a related article at [www.thinline.com/~ccoulomb/hunting.html].

505. Will, J. A. *The Case for the Use of Animals in Science*. Norwell, MA: Martinus Nijhoff, 1987. See related links at [www.osera.org/other2.htm].

506. Wolfensohn, Sarah, and Lloyd, Maggie. *Handbook of Laboratory Animal Management and Welfare*, 2nd ed. New York: Oxford Univ., 348 pp., 1998 (originally published in 1994). (0632050-527) This work is largely intended for British laboratories who are following regulations from the Home Office. See a pro vivisection article at [www.naiaonline.org/labanim.html].

507. Wolfson, David J. *Beyond the Law: Agribusiness and the Systemic Abuse of Animals Raised for Food or Food Production*. Glen Watkins, NY: Farm Sanctuary, 64 pp. The author is an attorney, who shows how most states have amended anti-cruelty laws so that farm animals have no legal pro-

tection. See a website demanding farm animal reforms at [www.farmusa.org].

508. Wood, Forrest E., Jr. *The Delights and Dilemmas of Hunting: The Hunting Versus Anti-Hunting Debate*. Lanham, MD: University Press of America, 252 pp., 1996. (0761804714) Read an article on attitudes to animals at [www.psyeta.org/sa/sa1.2/hills.html].

509. Woodcock, E. N. *Fifty Years a Hunter and Trapper*, reprint ed. Columbus, OH: A. R. Harding Publishing Co., 318 pp., 1991 (originally published in 1913). (0936622059) Read an anti fur article at [arrs.envirolink.org/ar-voices/ffa_fur.html]. A book like this probably could not be written now, due to the current state of wildlife populations and the lesser demand for fur. In the pioneering days of the nineteenth century, trapping was a fairly common profession. Unfortunately, it seems to have been so successful that they drove themselves out of business. How many full-time fur trappers are there in the United States today? Probably not many.

510. Woods, Geraldine. *Animal Experimentation & Testing: A Pro Con Issue*. Hot Pro Con Issues series. Springfield: Enslow, 1999. (0766011917) Written for juveniles.

511. Zayan, Rene. *Field Study into the International Transport of Animals and Field Study Concerning the Stunning of Slaughter Animals*. European Community Information Service, 1989. See an article by the meat industry on the humane stunning and slaughter of animals at [www.foodsafety.org/fs/fs088.htm].

512. Zinko, Ursula, et al. *From Guinea Pig to Computer Mouse: Alternative Methods for a Humane Education*. Winchester, England: Sarsen Press, 229 pp., 1997. A reference guide to about 400 alternative methods of animal education. Written by representatives of EuroNICHE, a Romanian animal welfare organization. See a EuroNICHE page at [www.euroniche.internetworking.de/EuroNiche/EuronicheHelp.html].

513. Zurlo, Joanne, et al. *Animals and Alternatives in Testing: History, Science and Ethics*. Larchmont, NY: Mary Ann Liebert, 86 pp., 1994. (0913113670) Zurlo works at the Johns Hopkins School of Public Health. Read this book online, beginning at [www.jhsph.edu/~caat/pubs/animal_alts/preface.htm].

Chapter 4

Nonfatal Uses of Animals

This chapter contains resources that discuss the human uses of animals that do not intentionally kill the animals. Because the creatures do not die, the controversy over these animal rights issues is less intense; but there are still some important moral questions to be considered. The common denominator in every nonfatal use of animals is their captivity. Some animal rights activists use anti slavery arguments and rhetoric to dispute the morality of keeping any creatures without their informed consent.

Animals are used in food production that does not kill, mainly for milk and eggs. Now it could be argued that milk production does end in death, since the milk cows are in fact killed when their milk output begins to decline. However, they do live for some years during their captivity. Opponents say that milk cows are injected with all sorts of hormones and chemicals to increase milk output, and that their lives are unpleasant or even painful.

One could also say that eggs are a fatal use of animals, as they are simply unborn animals; but by modern legal American definitions, neither eggs nor fetuses are often considered to be living creatures. The main problems in the egg-laying industry are the factory farming method of restrictive caging, with multiple hens trapped in tiny cages; debeaking, with many hens having their beaks cut off so that they will not peck each other; and the immediate choice of death for all hatched male chicks. Apparently male chicks are of no use since they do not grow fat enough to be valuable as poultry meat, nor are they useful as laying hens. They are generally dropped into plastic trash bags to suffocate, and are ground into pet food.

People also use animals to produce clothing, without sacrificing animals' lives in the process. The most common material of this type is wool. Sheep are sheared with clippers to remove much of their thick surface hair, which grows to keep them warm in the winter. Another type, far more expensive, and less well known, is silk. Insects such as silk worms are not often included in animal rights arguments, perhaps because they are repulsive to most people.

The classic example of animal captivity is that of the zoo or "menagerie." For many centuries, perhaps millennia, animals have been kept as money-making display items for entertainment purposes. People pay for the novelty of seeing unusual creatures, and these creatures are sometimes trained to act in human ways, like dancing bears or talking birds. Zoos have improved a great deal since the nineteenth-century norms of steel-barred cages. Nowadays they seek to simulate a real wild habitat, and this has had better results on the animals' psychology and the viewers' education. Zoos serve a multiple role of research, entertainment, education, and even conservation; several endangered species would be extinct if not for the captive breeding programs. Still, critics say that zoos perpetuate the myth of human superiority over animals.

Animals are used in formal therapeutic ways. The best example would be that of guide dogs for the blind, who assist blind people in navigating in public places, protecting them from unseen dangers. Dogs and other creatures are also used in psychological therapy, to help children or depressed people to "open up" emotionally.

The most widespread use of animals in captivity is that of keeping pets. Most American families have pets. The most common pets are dogs and cats, but birds, ferrets, turtles, fish, and more exotic types are also found. Because the practice of pet keeping is so common, there are a multitude of related problems.

First of all, the source of such pets is questionable. "Puppy mills" are where many pets are bred for public sale in pet shops. Many of these operations have been shown to be poorly managed with horrible conditions for the animals. On the other hand, there is a surplus of pets. Because owners do not keep their pets under close control, the animals often become pregnant and produce unwanted litters of puppies or kittens. These animals are often destroyed.

Pets do provide psychological, and thereby physical, health benefits. Elderly people are encouraged to have pets that give-and-take of love is an important part of life for lonely folks. However, some writers question this give-and-take, saying it is mostly take, since captivity is slavery. Domestication has in fact made many pet animals unfit for wild living. If we were to release these animals into "the wild," they would die. They do not have the instincts to survive well on their own any longer.

Animals are also used as transportation or as workers, mainly in rural and third-world countries. People use horses and oxen to pull plows and carts. The Pony Express was a well-known horse-and-rider combination in the Old West, delivering the mail to distant places.

The U.S. military and law enforcement agencies make use of animals. Police stations often have "K-9" units, whose duties include sniffing for drugs and explosives. The military has many more uses for animals. There have been messenger-carrying pigeons, mine-detection dogs, submarine-hunting doves, and torpedo-recovering dolphins. The space program has used primates in tests of orbital spacecraft.

There are a number of sports that make use of animals' special talents. Dogs, ferrets, and birds of prey are among those creatures which hunters sometimes use to help them catch game animals. There are also entertainment-style sports, like dog racing, horse racing, or rodeos.

Speaking of entertainment, many animal actors exist and become very famous. Perhaps the most famous of these actors would be Lassie (actually two collies) and Keiko ("Willy") the killer whale. Animal actors are used in television shows and cinematic movies quite frequently.

The most widely protested of the animal entertainment groups is the circus. There have been allegations of animal abuse in many circuses, though there have also been obvious improvements in the circus-animal programs.

Last, and perhaps most repugnant, is the rare use of animals known as bestiality. A small percentage of the world's population finds the practice of sexual relations with animals to be preferable to that with humans. This is illegal in most countries.

---

514. Acker, Duane, and Cunningham, Merle. *Animal Science and Industry: Principles of Successful Animal Production*, 5th ed. Englewood Cliffs, NJ: Prentice-Hall, 704 pp., 1998. (0135249015) The author was the president of Kansas State University and Assistant Secretary of Agriculture for Science and Education under President George Bush. Principles of successful domestic animal production are discussed, including nutrition, environment, marketing, and processing. The book touches upon the dairy, meat, wool, and poultry industries particularly. Well organized, with useful summaries. This is a textbook for college students in agricultural classes. Read the contents and a brief review at [www.prenticehall.ca/allbooks/ect_0135249015.html].

515. Amory, Cleveland. *Ranch of Dreams: The Heartwarming Story of America's Most Unusual Animal Sanctuary*. New York: Viking, 288 pp., 1997. (067087762x) Stories of Amory's adventures for animal protection, including abuses by circuses and the U.S. Navy. Until his recent death, he ran the Black Beauty Ranch in Texas. He was a very active animal rights proponent, founding The Fund for Animals, and helping to fund other causes like the Sea Shepherd Society. See the Black Beauty Ranch at [earthbase.org/home/essays/docs/black_beauty/index.html].

516. Anderson, Moira K. *Coping with the Sorrow of the Loss of Your Pet*, 2nd ed. Loveland, CO: Alpine Press, 188 pp., 1996. (0931866979) A variety of resources on this topic including a description of the stages of grief, methods of euthanasia, burial of deceased pets, helping children cope with the sadness, and websites with helpful information. See an online list of similar books at [www.petrelics.com/book2.htm]. There has been an explosion of books on this subject in recent years. Right or wrong, humans make pets into surrogate family members and often experience similar emotional reactions to the death of pets. For single people, pets are sometimes the only "family" available during everyday life.

517. Animal Transportation Association. *The Handbook of Live Animal Transport*. Fort Washington, MD: Silesia Companies, 1999. This looseleaf resource is updated periodically by subscription. See a summary of the contents at [www.lattmag.com/ handbook.htm]. This subject is largely neglected in animal rights discussions. The transportation of live animals is likely the source of much discomfort and death among animals. Trucks that carry livestock, for instance, are not generally climate controlled; the animals can overheat or freeze during inclement weather. Also, the forcing of a large number of creatures into a small space certainly creates an atmosphere of stress, with its related side effects of biting and kicking in the crowded environment.

518. Appleby, Michael. *Do Hens Suffer in Battery Cages?* Petersfield, England: Athene Trust, 1991. The author is a poultry expert from Glasgow University. He offers scientific evidence of hen suffering in factory farming environments. See more information on this subject at [worldanimal.net/ wan/hen-resources.html] and [www.berk.com/~alcreat/egg.html]. The poultry-industry argument that multiple hens are not affected by being trapped in wire-mesh cages seems ridiculous, on the face of it. If they were not disturbed by crowding, then wouldn't the common practices of debeaking and declawing be unnecessary?

519. Arkow, Phil, ed. *Pet Therapy: A Study and Resource Guide for the Use of Companion Animals in Selected Therapies*. Colorado Springs, CO: Humane Society of Pike's Peak, 196 pp., 1990. (0882477633) Read a thorough bibliography of pet therapy resources at [www.latham.org/ hcab.htm]. Pets can be extremely helpful in therapeutic environments. Most people perceive pets to be non threatening, and most pets are quick to give and receive affection from people.

520. Baker, Robert M., et al., eds. *Animals in Education: Value, Responsibilities and Questions*. Wellington, New Zealand: ANZCCART, 82 pp.,

1997. (064626379X) Proceedings of a 1996 conference. Read the table of contents at [www.adelaide.edu.au/ANZCCART/AnimalsEducation.html].

521. Beck, Alan M., and Katcher, A. H. *Between Pets and People: The Importance of Animal Companionship*, revised ed. New York: Putnam, 336 pp., 1996 (originally published in 1983). (0399127755) The value of pets and their impact on human lives. "Ultimately we concluded that to be healthy, it is necessary to make contact with other kinds of living things." (p. 9) Beck is a professor of animal ecology at Purdue University. Read articles at [text.nlm.nih.gov/nih/ta/www/03.html], [www.uclan.ac.uk/facs/health/soc work/swonweb/journal/issue1/pethum.htm] and [www.purdue.edu/UNS/html4ever/961108.Beck.book.html].

522. Bendiner, Robert. *The Fall of the Wild, the Rise of the Zoo*. New York: Dutton, 196 pp., 1981. (0525102701) It is a real possibility, which even famous naturalists like Jane Goodall have begun to admit, that some animal populations may not be able to survive in the wild, and they may only survive in captivity, in zoos. Some nations are unwilling or unable to save wild animal populations. Some animals exist only in small habitat areas, which are already heavily encroached upon by human populations. Is it morally acceptable then to capture animals from these endangered habitats and put them in zoos where they can be protected?

523. Benning, Lee Edwards. *The Pet Profiteers: The Exploitation of Pet Owners and Pets in America*. New York: Quadrangle, 211 pp., 1976. Many modern industries are profiting from American pet owners: kennels, pet shops, veterinarians, animal food makers, adoption agencies, and even pet cemeteries. Is this wrong? It seems to be a natural outflow of the traditional American capitalistic ideal: people love pets, demand for pet supplies and services are created, and industries meet the demand. Or has marketing artificially increased demand? To get an idea of the size of this industry, have a look at some of the various pet supply shops online at [dir.yahoo.com/Business_and_Economy/Companies/Animals/Supplies__Equipment__and_Gifts/].

524. Bloeme, Peter. *Frisbee Dogs: How to Raise, Train and Compete*, 2nd expanded ed. PRB Associates, 184 pp., 1994. (0962934623) The author has been training world champion frisbee dogs for more than 20 years. A good web page on frisbee dogs can be found at [www.spinningk9.com/]. Some animal rights proponents believe that any human use of an animal is exploitative in nature. When the animals participate in sports like frisbee catching, however, it seems that the dogs are delighted by the activity. Perhaps formal sporting events, like greyhound racing, have other

elements that would mitigate against the dog's enjoyment, but frisbee catching seems harmless enough.

525.Bostock, Stephen St. C. *Zoos and Animal Rights: The Ethics of Keeping Animals*. New York: Routledge, 232 pp., 1993. (0415050588) The author works for the Glasgow Zoo. He says that the rights of zoo animals are not necessarily degraded by captivity if those zoos provide a good quality of life. Zoos are important for conservation, science, and education. The book is reviewed at [www.psyeta.org/sa/sa2.2/marvin.html] and [www.nsplus.com/nsplus/insight/animal/captives.html].

526.Brestrup, Craig D. *Disposable Animals: Ending the Tragedy of Throwaway Pets*. Glen Rose, TX: Camino Bay, 240 pp., 1997. (0965728595) The author was the director of an animal shelter in Seattle. He challenges the practice of shelter euthanasia of animals, and says that owners must be more responsible. More than 10 million dogs and cats are killed yearly in the shelters. Read a good review at [www.animalpepl.org/97/5/books.html]. Read an excerpt from the book at [bestfriends.org/2001/2001brestrup.htm]. Some shelters have become "non killing." Is this reasonable and attainable? Some shelters run out of space; there are not enough homes for the animals there. In many cases, volunteers and workers from the shelters take more animals home than they may be able to care for. What is a solution to the sad truth of unwanted pets?

527.Briggs, Anna C. *Paws for Thought: How Animals Enrich Our Lives and How We Can Better Care for Them*. Leesburg, VA: National Humane Education Society, 128 pp., 1997. See the publisher's homepage at [nhes.org/]. One issue often addressed by the Humane Societies is that of "puppy mills," where most pet store puppies are born. See an article on this subject at [www.avar.org/pos_stat.htm#puppymills]. Another controversy is the relationship of local humane groups to the national ones. Often there is no affiliation. The national level humane organizations will often send out bulk mail, with the advertisements implying that donated funds will filter down to the local ones, as the following website disputes [mnvalley.pair.com/confused.htm]. Funds donated to national organizations are usually used for more advertisements and lobbying, not for local humane efforts.

528.Bryant, John. *Fettered Kingdoms: An Examination of a Changing Ethic*, revised 2nd ed. Arvada, CO: Fox, 84 pp., 1990 (originally published in 1982). This book is well known for its attack on the keeping of pets (slavery) and a controversial new epilogue with animal liberation tactics. "Let us allow the dog to disappear from our brick and concrete jungles — from our firesides, from the leather nooses and chains by which we enslave it." Read an article on this subject at [arrs.envirolink.org/ar-voices/pet_ slaves.html]. Are pets slaves? Do they have a choice? If you opened the door to let the pets choose to run in the wild or to live in your home, what would they do? In many cases, the domesticated animals could not survive in "the wild." They may not know how to hunt. Their physical characteristics have also been modified by centuries of human interference and intentional breeding, and they may not be even capable of living outdoors in inclement weather. But if the centuries of domestication are "immoral," are we justified in continuing the perpetuation of the helpless pet syndrome? It is no simple question.

529. Budiansky, Stephen. *The Covenant of the Wild: Why Animals Chose Domestication*. New Haven, CT: Yale University, 212 pp., 1999 (originally published in 1992). (0300079931) An interesting study of cultural perspectives on the evolution of domestication. The author says that evolution of genetics is more responsible for domestication than man's efforts. Domestic animals have reached a sort of symbiosis with the human race. He also disputes the animal rights position that says animals belong in the wild. Many animals have evolved to belong with man, and would not survive in the "wild." Read a related article at [www.xs4all.nl/~ianmacd/Jo/ast.htm].

530. Bulanda, Susan. *Ready! The Training of the Search and Rescue Dog*. Wilsonville, OR: Doral, 170 pp., 1994. (0944875416) A list of search-and-rescue dog resources can be found at [users.worldgate.com/~dognyard/booklist.html] and an article at [www.dogmania.com/articles/Rescue.html]. Many animals have senses far more accurate than those of humans. Dogs in particular are more sensitive to sounds and smells than people are. Perhaps the earliest human use of this canine trait was with "hunting dogs," but in recent times the practice has been extended to the hunting of humans: finding them under rubble or when lost in the countryside.

531. Burch, Mary R. *Volunteering with Your Pet: How to Get Involved in Animal Assisted Therapy with Any Kind of Pet*. Foster City, CA: IDG Books, 209 pp., 1996. (0876057911) Read about therapy dogs at [www.canismajor.com/dog/hmless.html]. Many therapeutic environments have been making use of animals in encouraging recovery among humans. Cats, dogs, birds, and fish are among the more commonly used creatures. People are much quicker to trust and share affection with animals than they would with human strangers.

532. Bustad, Leo K., and Spink, W. W. *Animals, Aging and the Aged*. Wesley W. Spink Lectures in Comparative Medicine, volume 5. Minneapolis: University of Minnesota, 227 pp., 1980. (0816609977) An article on use of pets for therapy with the elderly is found at [www.psyeta.org/sa/sa1.1/

perelle.html], and a lot of information about Bustad in his memorial and obituary is located at [members.aol.com/guyh7/Newsletter0998.htm]. Not all "retirement homes" recognize the value of animal contact with their clients. Instead they focus entirely on the cleanliness and sterility of their environments. There seems to be little interest on the emotional well-being of their patrons, only on the physical health, though physical and emotional health may not be so unrelated as we have been led to believe.

533. Castillo, Janet del, and Schwartz, Lois. *Backyard Racehorse: The Training Manual: A Comprehensive Off-Track Program for Owners and Training*, 3rd ed. Winter Haven, FL: Prediction Publications, 300 pp., 1997. (18844-75019) Find a bibliography of horse racing works at [www.horsenetwork.com/bookstore/booksrightracing.html]. See a brief anti race-horse site at [www.avar.org/pos_stat.htm#horseracing]. Is horse racing a moral problem? There have been a few gruesome cases of track accidents, leading to horse and human deaths. There have been cases of illegal drug injections to increase (or decrease) the horse's running ability. Any time money is involved, particularly the gambling done at most horse racing events, there will be criminal acts nearby. Is the training itself cruel, or is the mere fact that the horse is captive to human whims the problem?

534. Cherfas, Jeremy. *Zoo 2000: A Look Beyond the Bars*. London: British Broadcasting Company (BBC), 244 pp., 1984. (0563202815) Based on a British television series, discussing what the zoos of the future will look like. There has been a radical transformation of zoos over just the last hundred years. They have gone from beasts in small iron cages to centers of education and conservation. Still, debate rages over the entertainment aspect of zoos, the fact is, along with their lofty goals, they are in the business to make money as well. Are these goals incompatible? It seems to be working, with decent effects both for the animals and the businesses, in many cases.

535. Clutton-Brock, Juliet. *Domesticated Animals from Early Times*, 2nd ed. Austin: University of Texas, 208 pp., 1999 (originally published in 1981). Not yet published. The first edition appeared to be aimed at a young adult audience, so I assume the second edition may do the same. Read an interesting article about modern domestication of animals at [www.psyeta.org/sa/sa1.2/quinn.html].

536. Clutton-Brock, Juliet. *Horse Power: A History of the Horse and the Donkey in Human Societies*. Cambridge: Harvard University, 192 pp., 1992. (067440646x) The author is a zoologist from the Natural History Museum in London. Who can estimate the impact of horses and mules on human civilization? How long would the westward migration of Americans

have taken without them? Horses were used for transportation, military movements, pulling plows and other machinery, herding cattle, delivering mail, and much more.

537. Cochrane, Amanda, and Callen, Karena. *Dolphins and Their Power to Heal*. Rochester, NY: Healing Arts Press, 192 pp., 1992. (0892813865) See the table of contents at [www.gotoit.com/titles/dopohe.htm]. Read a related article at [whales.magna.com.au/POLICIES/levasseur/]. Dolphin therapy is an animal rights controversy for several reasons. The traditional medical community views this as another "holistic" or "Eastern" therapy, which means they believe it to be useless. Some hate the idea because we should not be keeping dolphins in tanks anyway; they should be free to roam at sea. Certainly there are some proponents of dolphin therapy who approach it from a very mystical viewpoint: that the porpoises somehow project healing through telepathy, or sonar waves, or whatever. Still, considering the good that comes from other "pet" encounters, why not find some good from dolphins as well?

538. Cohen, Robert, et al. *Milk: The Deadly Poison*. New York: Argus Archives, 317 pp., 1998. (0965919609) The author says that milk is full of fat and pus and viruses. He recently founded the AntiDairy Coalition, whose website you can see at [www.notmilk.com]. Reviewed at [www.montelis.com/satya/backissues/jul98/not_milk.html]. There are many articles and essays by prominent and scholarly authors, attacking the common American usage of milk as a drink. They say that half of the world's population is allergic (lactose intolerant) to milk. Animal rights proponents say that even aside from the human health angle, the dairy industry is cruel to its cows in the constant artificial insemination and use of enhancement drugs to keep the cows producing large amounts of milk. The milk industry is also where most veal calves come from, since dairy farmers have no use for male calves.

539. Collins, Miki, and Collins, Julie. *Dog Driver: A Guide for the Serious Musher*. Boulder, CO: Alpine, 386 pp., 1991. (0931866480) Dog sledding, particularly in Arctic races, is looked down upon by animal rights activists, who say the practice is cruel and dangerous for the dogs. A bibliography of sled dog resources can be found at [www.dogbooks.com/sled.html]. Long sled races like the famous Iditarod certainly do have their canine casualties, in greater proportion than horse racing, to be sure. On the other hand, animal rights activists want more than an end to dog sled races; they want an end to dogs on sleds, period. Is this reasonable to ask of native peoples? They have centuries-old traditions of using dogs to pull them around the Arctic Circle; it is really a necessity for their lives. Or is it? Why not use snowmobiles? Can they afford such machines, or the fuel?

540. Comfort, David. *First Pet History of the World.* New York: Fireside, 320 pp., 1994. (0671891022) How animals helped to shape human civilization. Read an article on pet ownership at [www.psyeta.org/sa/sa5.1/wells.html]. Studies have shown that pets have beneficial effects on people. Does this good effect justify the means of keeping animals in our homes? Some animal rights proponents call pet-keeping "slavery." This issue is complicated by the fact that domestic animals have been genetically altered by millennia of human usage and therefore probably could not survive in "the wild."

541. Cooper, Gale. *Animal People.* Boston: Houghton Mifflin, 205 pp., 1983. (0395321980) The author is a psychiatrist, who graduated from Harvard University. "Animal people are not just petkeepers . . . their relations with animals are of unusual intensity." (backcover) Have you ever met pet owners who go that extra mile in caring for their pets, who made you a little worried? I know a fellow who takes better care of his ferrets than most parents care for their children. Is that bad? There is no harm in it to the animals or to other people, that I can see. Perhaps there is a bit of unreality in the thinking of the owner, in projecting human needs on the animals. Or are the rest of us not recognizing those real needs in the animals?

542. Cooper, Jilly. *Animals in War.* London: Heinemann, 168 pp., 1983. (043-4143707) The role of animals in wartime, written by a best-selling British novelist. Is it right to use animals in war, in all types of duties? Is it all right for horses to carry the wounded away from a battle? How about pulling artillery to pound a city? Is it right to teach dolphins to find lost torpedoes? How about training them to attack and kill human swimmers who encroach on naval bases or threaten the ships? The issues become tricky, just as they are tricky for the humans who must also make these decisions for themselves. See a bibliography of the author at [billington.simplenet.com/JillyBooks.html].

543. Council for Science in Society. *Companion Animals in Society.* New York: Oxford, 78 pp., 1988. (0198576978) See a substantial bibliography on animals in society at [www.xs4all.nl/~ianmacd/Jo/halit.htm].

544. Croke, Vickie. *The Modern Ark: The Story of Zoos, Past, Present and Future.* New York: Scribner's, 320 pp., 1997. (0380731312) Why are zoos popular? What is the appeal? The author says that some zoos are good, but we should not try to replace wild habitats with zoos. Why shouldn't we? Aren't the animals safer in zoos? They will not be poached there, nor will they starve. Certainly we would like to have wild habitats, but which ones

are safe in this modern era? Or is this simply a human illusion? Are they really wild creatures, when set up safe and cozy in a fake environment?

545. Cropper, Katy. *One Woman and Her Dog: A Sheepdog's First Year* (video). London: Farming Press, 1997. Sixty-minute training video for the proper instruction of sheepdogs. See a bibliography of sheepdog resources at [www.bordercollie.org/bordbib.html].

546. Curtis, Patricia. *Dogs on the Case: Search Dogs Who Help Save Lives and Enforce the Law*. New York: Lodestar, 133 pp., 1989. (0525672745) Dogs used by the police, search and rescue, narcotics detectors, arson investigators, bomb sniffers. Written for young adults. Dogs are remarkably helpful to the police, when used and trained properly. They can locate missing people, find illegal drugs in luggage, sniff for explosives, and detect evidence of flammable agents at a fire scene. Does the fact that they are so good at such jobs mean that it is right for humans to use them for such jobs?

547. Cusack, Odean. *Pets and Mental Health*. New York: Haworth, 241 pp., 1996 (originally published in 1988). (0866568018) See an outstanding page of links about service and therapy dogs at [www.cofc.edu/~huntc/service.html] and a therapy dog training page at [doglogic.com/therapy.htm].

548. Davis, Karen. *Prisoned Chickens, Poisoned Eggs: An Inside Look at the Modern Poultry Industry*. Summertown, TN: Book Publishing, 175 pp., 1996. (1570670323) Includes both chickens and turkeys. Reviewed at [www.rasheit.org/VY-LIBRARY/libbkrevs.html#anchor1700080]. Read articles by the author at [www.upc-online.org/spring98/chicken_for_dinner.html], [www.montelis.com/satya/backissues/march97/fowl.html], and [arrs.envirolink.org/ar-voices/turkey.html]. The issue over eggs is substantially different from the issue over eating poultry meat. For one thing, the chicken does not die to give you the egg. Or does it? An egg is really a pre-baby chicken. It is a similar question to the human questions over abortion: When is the fetus a human life? When is a chick in an egg a chicken life? Or does it even matter? Does it matter that the hen may not have a semblance of a decent life? Is that irrelevant, so long as the hen lives? And does it matter if the hen is killed immediately upon a decline in egg-laying productivity? These are some of the questions arising from the egg industry.

549. Davis, Kathy Diamond. *Therapy Dogs: Training Your Dog to Reach Others*. New York: Howell, 212 pp., 1992. (0876057768) See an article on

federal policies regarding access in public buildings for service animals at [www.sonic.net/~melissk/svc_anim.html].

550. Dawne, Diana. *Venture's Story: Life & Times of a Guide Dog*. Orange, CA: Word & Pictures Press, 285 pp., 1997. (0964485737) Written by Venture's owner, on training a Labrador Retriever to become a guide dog. Read an excerpt at [www.light-communications.com/author/dawne/sample.htm], and see the author's homepage at [www.yumston.com/book.html]. Here is one of the tougher questions for anti pet activists: Is it wrong for a blind person to keep a dog to help him or her get around? Maybe in the future, robotics and computers will be advanced enough to eliminate the need for such animal help. But would the person prefer a robot to an animal?

551. Dekkers, Midas. *Dearest Pet: On Bestiality*. New York: Norton, 208 pp., 1994. (0860914623) Read a brief explanation of bestiality at [cfs.he.utk.edu/240class/atypical.htm]. This may be the most controversial nonfatal use of animals because the vast majority of people would condemn sex with nonhuman animals as being wrong (animal rights or no). Is it a harmless outlet for sexual urges, or is it simply perversion? Most states and countries consider bestiality to be illegal, though enforcement is rare.

552. Department of the Army. *Packing: A Guide to the Care, Training, and Use of Horses and Mules as Pack Animals*. Flagstaff, AZ: Northland, 102 pp., 1989. This is probably a reprint of World War I and II-era procedures, when animals (not mechanized vehicles) were a common method of military transport. See a good related website at [www.qmfound.com/horse.htm].

553. Domalain, Jean-Yves. *The Animal Connection: The Confessions of an Ex-Wild Animal Trafficker*. New York: Morrow, 250 pp., 1977. (0688031692) This author had a fascinating life: he was a headhunter in Borneo. Later he began to export animals illegally in Southeast Asia. After four years, he realized the error of his ways, ended his business, and wrote this book. He says that 80 percent of the illegally obtained animals die in transit or within months of arrival. See the Canadian laws on wild animal trading at [www.nearctica.com/conserve/canlaw/act_e.htm]. This is a vicious industry by almost any account, where whole family groups of animals are destroyed to steal the babies for sale to zoos and pet shops. Because many animals are protective of their young, the poachers must usually kill the parents and other adults to get the young ones. The transport methods are necessarily poor (in attempts to hide the animals from authorities). Between the shock of instant family massacre, captivity, and poor transport provision, the animals often become sick and die.

554. Donoghue, Michael, and Wheeler, Annie. *Save the Dolphins.* Toronto: Key Porter, 119 pp., 1990. (0924486074) Names the many threats to dolphins: pollution, driftnetting, and captivity. In recent years, public outrage about the "incidental kills" of dolphins in giant tuna nets led to major changes in the tuna industry. Why are dolphins so popular? They were popular even before any of the modern studies on their intelligence. Is it because they look so friendly, with that smile-shaped look permanently on its snout? Is it true, as critics often say, that animal rights people only care about the cute and cuddly creatures? It is certainly easier to raise funds to defend the cute and cuddly ones.

555. Drakeford, Jackie. *Working Ferrets: Selection, Training and Sport.* Shrewsbury, England: Swan Hill Press, 128 pp., 1996. (1853108049) Preparing ferrets to chase rabbits from their holes and burrows. Includes tangential topics, like choosing a gun and preparing nets for the rabbit capture. Ferreting has been a sport for many centuries, mainly practiced in continental Europe.

556. Drayer, Mary Ellen, ed. *The Animal Dealers: Evidence of Abuse of Animals in the Commercial Trade, 1952 - 1997.* Washington, DC: Animal Welfare Institute, 341 pp., 1998. Addresses the issues of pet dealers, and those who sell animals to laboratories. See the introduction at [www.animalwelfare.com/pubs/bk-conts.htm#dealers] and a brief publisher review at [www.geocities.com/Heartland/Plains/8677/expose.html]. The issue of Class B licenses issued by the USDA, which allow the collection of animals from "miscellaneous" sources (accounting for 20% of lab animals acquired), has become very prominent recently. Critics say that unscrupulous dealers actually steal pets from homes and sell them to out-of-state labs for testing. Opponents say that this is an "urban legend" and rarely occurs. Some states are now considering legislation to make Class B licensing illegal.

557. Druce, Clare. *Chicken and Egg: Who Pays the Price?* London: Green Print, 1989. (1854250280) A criticism of the poultry industry; claims that many chickens are diseased and pass on diseases to humans. Read a brief review at [www.mcspotlight.org/media/books/druce.html]. The conditions in poultry factory farms are generally appalling. Human workers grow quite sick with short trips inside, often requiring breathing apparatus. That many birds with that much guano waste produce dust and ammonia, which fill the air. The caging of the birds has made cleaning easier (waste falls to the ground), but there is some question as to whether owners have actually improved the atmospheres inside the buildings.

558.Duet, Karen Freeman. *The Business Security K-9! Selection and Training.* Foster City, CA: IDG Books, 260 pp., 1995. (0876054394) Helpful information on general breed temperaments and legal ramifications of guard dog use. See an excellent list of related books and videos at [www.adognet.com/4M/guarddog.html]. Guard dogs are used by many businesses to patrol and protect large grounds. Guard dogs are much cheaper than human guards over the long term. However, guard dogs are also far less capable of making flexible decisions. In general, they would act noisily to deter illegal entrance and violently to stop trespassers. There is no negotiation or thoughtful response to situations in guard dogs, as there could be in a human guard.

559.Durning, Alan, and Brough, Holly. *Taking Stock: Animal Farming and the Environment.* Worldwatch Paper #103. Washington, DC: Worldwatch Institute, 70 pp., 1991. (1878071041) The environmental cost of animal farming. See an article on factory farming at [www.avar.org/pos_stat.htm#factory-farming]. Recent studies have shown a major environmental impact of factory farming on the land. That heavy population density of animals leads to a number of problems, including sewage run-off.

560.Eames, Ed, et al. *Partners in Independence: A Success Story of Dogs and the Disabled.* New York: Howell, 227 pp., 1997. (0876055951) Eames is blind and uses a dog for assistance. The book discusses dogs used for the blind, deaf, and physically disabled.

561.Evans, Mark. *Pet Rescue.* London: Fourth Estate, 128 pp., 1997. (07531-54463) The RSPCA (Royal Society for the Prevention of Cruelty to Animals) on subjects of animal care, racing of greyhound dogs, depressed pets, and abused dogs. The book is based on a popular British television series. The old adage of the family progression of violence — husband yells at wife, wife yells at kids, kids kick the dog — is often true. It is also known that animal abuse is often a precursor of violence against people. Yet animal abuse is rarely enforced and even more rarely penalized. See information of the Federal Emergency Management Agency (FEMA) about helping animals during localized disasters, like floods, at [www.fema.gov/fema/iwrcpap.htm].

562.Fogle, Bruce. *People & Pets: A Guide to Choosing, Understanding and Caring for Your Pet.* London: Boxtree, 1990 (originally published in 1984). Why do we keep pets? This is a good question. Do we keep pets to prove our power over "lower" species? Is it the proof of a patriarchal culture as the feminists claim? Is it an unconscious realization of the emotional health benefits that may be gained by having pets? Or is it just a tradition; our parents did it, so we do it too?

563. Fox, Michael W. *Beyond Evolution: The Genetically Altered Future of Plants, Animals, the Earth . . . and Humans.* New York: Lyons, 256 pp., 1999. (1558219013) Sequel to *Superpigs and Wondercorn.* Fox issues major cautions on our future, due to genetic engineering attempts. Read a brief article by the author at [www.natural-law.ca/genetic/NewsMay-June99/GEN5-18MFoxHumaneSoc.html]. One of the worrisome aspects of genetic engineering is the hidden nature of long-term effects. Perhaps the addition or subtraction of gene Z produces useful changes in today's creatures, but will it produce horrific results in future generations? How long should we do testing before accepting genetic alterations as being safe? And how do we test for such things — with animals, or on people?

564. Fox, Michael W. *Superpigs and Wondercorn: The Brave New World of Biotechnology and Where it All May Lead.* New York: Lyons & Burford, 209 pp., 1992. (1558211829) The author calls bioengineering "genetic imperialism" and "biological fascism." He lists dangerous and ridiculous experiments in genetic engineering, like 12-foot-long pigs and cows weighing five tons. See a good article on the dangers of biotechnology at [www.purefood.org/eight.html]. Once the decision has been made that animals can be rightly used for food, is there any reason not to proceed with genetically engineering fatter animals? How about creatures so fat that they cannot walk? How about pigs with no legs? If they are only food machines, there is no need for them to run around, right? Or will we simply militate against such things because of the possible dangers of genetic engineering?

565. Fraser, A. F., and Broom, D. M. *Farm Animal Behaviour and Welfare*, 3rd ed. reprint. Wallingford, England: CAB International, 437 pp., 1997 (originally published in 1974). (0851991602) "This book . . . reviews the scientific information which is available concerning the assessment of animal welfare, and the evaluation of the effects on animals of different management methods and housing conditions." (p. vii) "Behaviour has always been recognized as an important indicator of health and welfare in farm animals." (backcover)

566. Genders, Roy. *The National Greyhound Racing Club Book of Greyhound Racing: The Complete History of the Sport*, reprint ed. London: Pelham, 400 pp., 1990 (originally published in 1981). (0720711061) See a site about sporting dogs at [www.canismajor.com/dog/hunttrl.html]. Read an anti-greyhound racing site at [www.ocpa.net/news_1_21_99_greyhounds.htm]. One of the major claims against greyhound racing is the short-term nature of the sport. Most animals are good in racing for only a short time, and then the animals are destroyed. Recently a movement has arisen within the sport to offer the greyhounds for adoption, and many have found

homes. Still, is it right to create an animal for a solely entertainment purpose (gambling) and then abandon it? Isn't this irresponsible? Or is it no different than raising a cow for milk? Once its production slips, the cow is killed.

567. Grandin, Temple. *Livestock Handling and Transport.* Wallingford, England: CAB International, 352 pp., 1993. (0851988555) Read an article by the author at [www.montelis.com/satya/backissues/jan98/straight.html]. See a related bibliography at [www.nal.usda.org/awic/pubs/oldbib/qb9432.htm]. See the author's impressive home page with animal-issues links at [www.grandin.com].

568. Haggerty, Arthur J. *How to Get Your Pet into Show Business.* New York: Howell, 194 pp., 1994. (0876055595) Read a brief article by the author at [www.lovemypets.com/love/haggerty.htm] and a site about the author at [www.lovemypets.com/articles/haggerty.htm]. Read a relevant bibliography at [www.altpet.net/book/actors.html]. Animals on television and in the movies are very popular. Does the educational or emotional good brought to television audiences justify their use as actors, or does such acting simply reinforce the human idea that animals are "lower" and we can do with them what we wish? Does animal acting reinforce anthropomorphic ideas, that the animals are just like us (when that is a debatable point)?

569. Harlow, Mike. *K-9 Bodyguards.* Neptune City, NJ: TFH, 160 pp., 1995. (079-3802873) See the author's homepage at [www.geocities.com/SoHo/8707/index.html]. Read an article on dogs in police work at [k9tactics.com/special.htm].

570. Harris, Lee William. *Falconry for Beginners: An Introduction to the Sport.* Shrewsbury: Swan Hill Press, 160 pp., 1999. (1853108936) A training manual. See a good falconry page at [www.americanfalconry.com/FirstYear.html] and a falcon book list at [www.users.zetnet.co.uk/omorgan/books/falcon.html]. Falcons are being used in ways beyond simple hunting sports. In many major cities, pigeons have become overpopulated, as they scavenge on human waste. Some cities have intentionally installed falcon families and nesting sites on the skyscrapers so that this natural predator can reduce pigeon populations.

571. Hearne, Vicki. *Adam's Task: Calling Animals by Name*, 5th ed. New York: Harper Perennial, 274 pp., 1994 (originally published in 1982). (03945-42142) The author is a professional dog trainer and an assistant professor of philosophy at Yale University. She attempts to define why and how animals are trained.

572.Helfer, Ralph. *Beauty of the Beasts: Tales of Hollywood's Wild Animal Stars*. Los Angeles: Tarcher, 223 pp., 1990. (0874775167) Some activists say that wild animals are degraded when trained to do unnatural acts. Dancing bears? Elephants doing pirouettes? Zookeepers say that they teach elephants tricks not entirely for entertainment purposes but because elephants become bored and need to learn new things. What sort of life does the animal actor have when not in the movie? Is its habitat interesting and proper, or does it live in an iron cage? Does it have a social life or proper medical care? There have been improvements in animal treatment on movie sets (note the common movie-credit caveat "no animals were harmed during the making of this film").

573.Hess, Elizabeth. *Lost and Found: Dogs, Cats, and Everyday Heroes at a Country Animal Shelter*. New York: Harcourt Brace, 240 pp., 1998. (015-1003378) Good, bad, and inexplicable characters you would meet at your average animal shelter. Reviewed at [vh1459.infi.net/living/docs/petdoc05 2698.htm] and [www.montelis.com/satya/backissues/sep98/shelter.html]. Read some counter claims to the overpopulation problem at [www. naia.online.org/overpop1.html]. See statistics on animal shelters at [certi ficate.net/wwio/pet26.htm].

574.Hoage, R. J., and Deiss, William A., eds. *New Worlds, New Animals: From Menagerie to Zoological Park in the Nineteenth Century*. Baltimore, MD: Johns Hopkins University, 198 pp., 1996. (0801851106) The development of zoos from sideshows to conservation and education centers. Based on papers from a 1996 symposium. See one zoo history site at [www.zoo.org/ history/history1.html].

575.Hooper, Frederick. *The Military Horse: The Equestrian Warrior Through the Ages*. New York: Barnes, 105 pp., 1976. (0498019926) See a short bibliography at [www.horsebooks.co.uk/c12.htm] and a military horse site at [www.militaryhorse.org/]. Military horses require more training than farm horses. Conditioning animals not to be afraid (and run away) from gunshots and artillery explosions is a difficult process. This is why elephants did not often do well as cavalry weapons in ancient warfare: they were very prone to panic. Horses are more easily controlled in battle.

576.Hulse, Virgil M. *Mad Cows and Milk Gate*. Phoenix, AZ: Marble Mountain, 338 pp., 1996. (0965437701) The author says that most American dairy cows have bovine leukemia, bovine tuberculosis, and the bovine immunodeficiency virus. The author was a milk and dairy inspector in California for 13 years; now he is a Fellow at the American College of Preventative Medicine. Mad Cow Disease is one of the most interesting and frightening of the modern medical conditions, and it is not really a disease. It appears to have come from cannibalism, but began to affect our industrial nations with the agricultural practice of using slaughterhouse scraps of brain in with the other refuse to mix into cattle feed for protein. The corrupted brain matter may later trigger corruptions in the other animals that are eating it. It can also spread between species, cows to humans, for example. It led Great Britain to destroy millions of cows when many humans died of the infection.

577.Irving, Jow. *Gundogs: Their Learning Chain*, 2nd ed. Shrewsbury, England: Swan Hill Press, 205 pp., 1999. (1840370335) On training dogs to be hunting companions. See a good bibliography on dogs in sports at [www.users.zetnet.co.uk/omorgan/books/dogs.html]. Is it immoral to teach an animal to kill? In the Bible (and often in United States courtrooms), the owner of an animal is responsible for the animal's actions.

578.Johnson, William. *The Rose-Tinted Menagerie*. Washington, DC: People for the Ethical Treatment of Animals (PeTA), 335 pp., 1991. (09460-97283) An expose about zoos, circuses, and aquariums, from behind the scenes. Read the whole book at [www.iridescent_publishing.com]. See a pro circus page at [www.optonline.com/comptons/ceo/01007_A.html] and an anti circus page at [circuses.com/cindex.html].

579.Joys, Joanne Carol. *The Wild Animal Trainer in America*. Boulder, CO: Pruett, 327 pp., 1983. (0871086212) Stories of famous animal trainers; many photographs. It is interesting how different our views of animal trainers may be, based solely on the type of animals being trained. Many people value a good dog trainer or horse trainer, when the training is leading to the improvement of an animal as a happy pet. But the animal rights movement has led to the demonization of wild animal trainers, since these trainers are working with animals that are believed to be unethically captured and used. See an article about becoming a wild animal trainer at [www.arkanimals.com/Career/Career7.html].

580.Kennedy, Malcolm J. *Hauling the Loads: A History of Australia's Working Horses and Bullocks*. Melbourne, Australia: Melbourne University, 240 pp., 1992. (0522844553) Briefly reviewed at [www.users.bigpond.com/Oldavo/reviews.htm].

581.Kete, Kathleen. *The Beast in the Boudoir: Pet-Keeping in Nineteenth Century Paris*. Berkeley: University of California, 210 pp., 1995. (05200-71018) This book is based on a doctoral thesis. The premise is that pet care is an example or microcosm of a general worldview shift in the nineteenth century, when the viewed value of pets changed from worthlessness to pricelessness. An unusual related site can be found at [www.thepetvine.com/group/DogsnCatsThroughHistory].

582.Kiley-Worthington, Marthe. *Animals in Circuses and Zoos — Chiron's World?* Basildon, England: Little Eco Farms, 1990. This study, which was funded by the United Kingdom RSPCA, investigates animal treatment in circuses. It cites "evidence of considerable suffering and very poor housing amongst all circus animals." Read a summary at [www.rspca.org.au/pack.html]. One major factor in the controversy over circus animals is that of transportation. Circuses tend to travel from city to city frequently, and the transport of animals by train or truck is not simple. Are the vehicles climate controlled? Do the animals receive proper food, water, and medical treatment as necessary? What kind of stresses are the animals feeling in crowded conditions during transport?

583.Kittrell, William, et al., eds. *Pet Therapy: An Anthology*. Healing Arts and Sciences series, volume 6. Virginia Beach, VA: Grunwald & Radcliff Publishers, 1992. (0915133385) Go to a nice site about pet therapy at [www.geneseo.edu/~ece1/pet.html].

584.Kneidel, Sally Stenhouse. *Classroom Critters and the Scientific Method*. Golden, CO: Fulcrum, 176 pp., 1999. (1555919693) A list of 37 quick (and nonfatal) experiments with hamsters, goldfish, lizards, and other small animals to show classroom kids. Read excerpts at [fulcrum-resources.com/html/ classroom_critters.html]. A lot can be learned from animals without dissecting them. Modern biologists have begun to notice this, as naturalists gather data from animals in the wild (e.g., Jane Goodall and Dian Fossey) rather than poking and prodding them in laboratories.

585.Koebner, Linda. *Zoo Book: The Evolution of Wildlife Conservation Centers*. New York: Tom Doherty Associates, 192 pp., 1994. (03128-5322x) "[T]he ways zoos were and the way they are going to be." This is a large folio-sized book with many photographs. See a site with several articles written by zoo personnel on conservation efforts at [www.selu.com/bio/cauz/papers/index.html].

586.Kramer, Mark. *Three Farms: Making Milk, Meat, and Money from the American Soil*. Boston: Little, Brown, 256 pp., 1986. (0674889363) Read a good article on meat and milk production ethics at [www.astridlindgren.com/english/ethical.htm].

587.Lagoni, Laurel S., Hetts, Suzanne, and Butler, Carolyn, et al., eds. *The Human-Animal Bond and Grief*. Philadelphia: Saunders, 496 pp., 1994. (0721645771) See a good article on this topic at [www.psyeta.org/sa/sa4.2/

stephens.html]. Thinking about this from another angle: how do the animals feel when their human "owners" die, or when fellow pets die? Many creatures do seem to grieve. There has been a growing industry of pet psychologists. I am not very familiar with their work and cannot say whether pet psychologists could help an animal. I can say that my own pets do seem to adore each other, and I believe they would feel grief if one were to disappear suddenly.

588.Lasher, Margot. *And the Animals Will Teach You: Discovering Ourselves Through Our Relationships with Animals.* New York: Berkley, 240 pp., 1996. (0425154580) You can even get a college degree in this sort of human/animal relationships field. See one example at [www.soton.ac.uk/~azi/azi.htm].

589.Lawrence, Elizabeth Atwood. *Hoofbeats and Society: Studies of Human-Horse Interactions.* Bloomington: Indiana University, 1985. (0253328438) Read an article by the author at [www.psyeta.org/sa/sa2.2/lawrence.html]. Horses can be very intelligent creatures. Do they "mind" that humans appropriate them for riding or pulling carts? In my experience, they seem to mind use by some folks but relish it with other humans, just as pets develop preferences for certain family members over others.

590.Lawrence, Elizabeth Atwood. *Rodeo: An Anthropologist Looks at the Wild and the Tame*, reprint ed. Knoxville: University of Tennessee, 288 pp., 1984 (originally published in 1982). (0226469557) A good rodeo site is found at [www.imh.org/nchf/rodeo2.html]. One of the controversial aspects of rodeos among animal rights activists is that of the bucking events, both broncos and bulls. Usually the animal is "encouraged" to buck by use of a strap that is cinched around the animal tightly. Rodeo people say the strap is annoying but not painful, while opponents claim it causes a lot of pain to the animal. Either way, it does tend to make the animals buck and thus give the cowboys a more violent ride to the delight of the crowds.

591.Lemish, Michael G. *War Dogs: Canines in Combat.* New York: Brassey's, 224 pp., 1996. (1574880179) Dogs have had a wide number of uses in combat situations, such as finding booby traps and mines, doing patrols and reconnaissance, pulling sleds across snow, and guarding certain zones from intruders. See a bibliography of war dogs at [www.qmfound.com/War_Dogs_Bibliography.htm].

592.Levinson, Boris M., and Mallon, Gerald P. *Pet-Oriented Psychotherapy*, 2nd ed. revised and expanded. Springfield, IL: Charles C. Thomas, 242 pp., 1996. (0398066736) Related links at [trfn.clpgh.org/animalfriends/therapy.html].

593.Lewis, George, and Fish, Byron. *I Loved Rogues: The Life of an Elephant Tramp*, revised ed. Seattle: Superior Publishing Co., 184 pp., 1978 (originally published in 1955). (0875649548) The amazing life and stories of a circus elephant trainer, with numerous photographs. While zoos and circuses have surely improved, this book makes clear that most captive elephants die from gunshot wounds for misbehavior. See an anti circus site at [circuses.com/ringling/ele3.html]. One disturbing thing that I learned from this book was the constant flux of elephant trainers moving in and out of each circus. They rarely stayed more than a year in one place. Elephants are very social and loyal creatures, and I wonder if many of their behavioral problems do not stem from the constant adjustment to new "bosses."

594.Liebhardt, William C., ed. *The Dairy Debate: Consequences of Bovine Growth Hormone and Rotational Grazing Technologies.* Davis: University of California at Davis, 372 pp., 1993? Briefly reviewed at [www.animal welfare.com/farm/bkrviews.htm].

595.Luallin, Carol S. *Animal-Assisted Therapy: An Annotated Bibliography (1990-1996).* Emporia, KS: Emporia State University, 1996. Read the bibliography at [www.swcp.com/~ldraper/slim/biblios/luallin.html]. See a site on dogs used to help the disabled at [www.dfd.org.uk/RRG.HTM].

596.Lufkin, Elise. *Found Dogs: Tales of Strays Who Landed on Their Feet.* New York: Howell, 127 pp., 1997. (0876055978) Stories of successful adoptions of stray dogs from animal shelters. Two of the adopters in the book are Al Gore and Jamie Lee Curtis. See the author's website at [www.founddogs.com/html/more.html]. Read about some of the problems that can arise with stray adoptees at [www.naiaonline.org/resc95gc.html]. The best dog my family ever owned was obtained from a shelter. Many people mistakenly believe that shelter dogs must be there for behavioral problems ("they bite"), when in fact many or most animals are simply lost or their owners abandoned them during a move. Furthermore, the shelters tend to screen the animals for diseases, give them vaccinations, and provide thorough checkups to ensure that adopters will find the animals to be healthy.

597.Luoma, J. R. *Crowded Ark: The Role of Zoos in Wildlife Conservation.* Boston: Houghton Mifflin, 209 pp., 1987. (0395408792) The author is a contributing editor for *Audubon* magazine. A list of books recommended for zoos can be read at [www.sil.si.edu/SILPublications/zoo-aquarium/i-vii. htm]. Modern zoos are not simply passive helpers of animal populations by keeping a few endangered creatures safe and warm. Many zoos participate with each other in breeding programs, so that the populations do not become too inbred and to increase their numbers. Also, specimens are kept of the animals' reproductive fluids so that critically endangered species might be brought back in the future using advanced genetic processes. Of course, the education of the public provided by the zoo, regarding the endangerment of a species, is a great help to the cause of protecting wild habitats.

598.Lutz, Don. *The Weaning of America: The Case against Dairy Products . . . and 12 Other Essays on Ethics, Animals and Planetary Survival.* Atlanta, GA: Inner Peace Books, 1997. (0963027514) "Why are humans the only animals that refused to be weaned?" Reviewed at [www.ahimsaworks.com].

599.Malamud, Randy. *Reading Zoos: Representations of Animals and Captivity.* New York: New York University, 377 pp., 1998. (0814756034) The author is an English professor at Georgia State University. He says that zoos teach that "we are, by nature, an imperial species: that our power and ingenuity entitles us to violate the natural order by tearing animals from the fabric of their ecosystems and displaying them in an 'order' of our own making." (backcover) Zoos misrepresent animals' existences and reinforce the validity of trapping and controlling them. This view may be true about the more common species, like kangaroos, which are not endangered in the wild. How about with endangered species? Is the only proper way to view a wild animal to actually travel to the wild? Must we go to Australia to see a real kangaroo?

600.Maple, Terry L., and Archibald, Erika F. *Zoo Man: Inside and Outside the Cages.* Marietta, GA: Longstreet, 196 pp., 1993. (1563520168) The author is the director of the Atlanta Zoo and a psychology professor at Georgia Institute of Technology. He has been very successful in upgrading this zoo. See his web page at [www.zooatlanta.otg/html/maple.html]. The book is reviewed at [www.psyeta.org/sa/sa2.2/marvin.html]. Zoo work, just like any other job, has unrelated aspects that can be troublesome. I worked as a zoo volunteer and discovered that there are politics and poor planning, as in any organization.

601.Matthews, Mark. *The Horseman: Obsessions of a Zoophile.* New Concepts in Human Sexuality series. Buffalo, NY: Prometheus, 208 pp., 1994. (087-975902x) The emotional and moral issues of bestiality. Resources on the topic at [www.amba.demon.co.uk/sex/bestres.htm]. As mentioned earlier, this nonfatal use of animals is largely frowned upon by societies and is usually illegal. In some cases it may also be dangerous, as some diseases of animals may spread to the human, or the human may spread a disease to the animal. Interspecies contact is one of the suspects for the AIDs virus, as

African primates seem to have the disease yet are not affected, while it is fatal in humans.

602. McElroy, Susan Chernak. *Animals as Teachers and Healers: True Stories and Reflections*. New York: Ballantine, 178 pp., 1997. (0345421175) Ten years after her diagnosis of cancer, the author tells how animals helped her to survive. Read an article by the author at [www.deltasociety.org/dsm000.htm] and a brief review at [www.aspca.org/calendar/watrevw6.htm].

603. McKenna, Virginia, Travers, William, and Wray, Jonathan, eds. *Beyond the Bars: The Zoo Dilemma*. Wellingsborough, England: Thorsons, 208 pp., 1988. (0722513631) The authors show the immorality of animal captivity. Read an anti zoo article at [www.nonline.com/procon/html/con Zoos.htm]. Interestingly enough, many of the barriers encountered at zoos are not to keep the animals trapped inside so much as to keep the people away from the animals. Humans, either out of stupidity, curiosity, or maliciousness, often climb in to "play" with the beasts. In some cases this is a fatal error; in other cases the animals have protected the human from harm.

604. McMains, Joel M. *Manstopper! Training a Canine Guardian*. Foster City, CA: IDG Books, 244 pp., 1998. (0876051441) A former police officer on how to train guard dogs. Training methods are an issue of debate. The old "negative reinforcement" theory says that when the animal errs, one should hit it or frighten it so it will not repeat the error. The newer "positive reinforcement" teaches that rewards are more effective than punishments. It seems that a combination works best (though that is a utilitarian justification; results do not prove that the method is right).

605. Milani, Myrna M. *Preparing for the Loss of Your Pet: Saying Goodbye with Love, Dignity, and Peace of Mind*. Rocklin, CA: Prima, 377 pp., 1998. (0761516484) A veterinarian talks about pet owner preparations for the death of a pet, how to talk to children about it, and what sort of emotional responses to expect from yourself.

606. Morgan, Paul B. *K-9 Soldiers: Vietnam and After*. Central Point, OR: Hellgate Press, 184 pp., 1999. (1555714951) Stories told by an Army major. After the war he started a security business with K-9 units. See a memorial of the K-9 military dogs killed in action in Vietnam at [www.vdhaonline.org/dogskia.htm]. Read an excerpt on the Amazon.com book pages.

607. Myers, Olin Eugene. *Children and Animals: Social Development and Our Connections to Other Species*. Lives in Context series. Boulder, CO:

Westview, 208 pp., 1998. (0813331714) James Serpell writes that "This book presents a radical reappraisal of the importance of animals in human development. . . . Myers posits that animal-human interaction plays a uniquely significant role in the growth of preverbal self-awareness and moral sense." The importance of animals in child development, and how children begin to develop a sense of self through differentiation with animals. Read an article by the author at [www.psyeta.org/sa/sa4.1/myers.html].

608. National Research Council. *The Psychological Well-Being of Non-human Primates*. Washington, DC: National Academy Press, 168 pp., 1998. (0309052335) How to meet the requirements of the 1985 Amendment to the Animal Welfare Act, to promote the emotional health of primates. Housing and sanitation, daily care routines, and social companionship needs are addressed. Read the whole book at [www.nap.edu].

609. Nichol, John. *The Animal Smugglers: and Other Wildlife Traders*. New York: Facts on File, 196 pp., 1987. (0792444124) Analysis of the illegal animal trade: it's background, it's scale, and specific cases. Read related articles at [www.envirolink.org/archives/enews/0453.html] and [arrs.envirolink.org/ ar-voices/exotic.html].

610. Nilsson, Greta. *The Bird Business: a Study of the Commercial Cage Bird Trade*, 2nd ed. Washington, DC: Animal Welfare Institute, 121 pp., 1981 (originally published in 1977). See a brief statement on the bird trade at [www.avar.org/pos_stat.htm#wildbirds]. One of the questions specifically related to the keeping of birds in cages is the obvious limitation of their freedom. Unlike other pets, these birds are generally kept forever in rather tight confines, unable to even fly. Some owners will use scissors to "clip their wings" so that they can come out occasionally (the bird can only fly for short distances), but not all have even this freedom.

611. Norton, Bryan G., et al. *Ethics on the Ark: Zoos, Animal Welfare, and Wildlife Conservation*. Washington, DC: Smithsonian, 330 pp., 1995. (1560986891) All sides of the zoo questions, pro and con. The author is a philosophy professor at Georgia Institute of Technology. See a brief author statement at [198.240.72.81/case11.html].

612. O'Barry, Richard, et al. *Behind the Dolphin Smile*. Renaissance Books, 288 pp., 1999 (originally published in 1988). (1580631010) The author was the trainer of "Flipper" from the popular television series. This book is an analysis of the issue of keeping dolphins in captivity; the author concludes that dolphins should be free. Reviewed at [www.physics.helsinki.fi/whale/intersp/pages/book1.htm]. Why do dolphins, whales, and elephants

get such special attention for freedom-from-captivity issues? Because they are large, or because they are probably the most intelligent creatures in captivity?

613. Olsen, Sandra L., ed. *Horses Through Time.* Niwot: Roberts Rinehart, 222 pp., 1996. (1570980608) Compiled essays by ten contributors on horses being hunted, domesticated, used for medicine, ridden, and trained for sports.

614. Oski, Frank A. *Don't Drink Your Milk! New Frightening Medical Facts about the World's Most Overrated Nutrient*, 9th ed. Brushton, NY: Teach Services, 96 pp., 1992 (originally published in 1983). (0945383347) "The fact is: the drinking of cow milk has been linked to iron-deficiency anemia in infants and children; it has been named as the cause of cramps and diarrhea in much of the world's population, and the cause of multiple forms of allergy as well; and the possibility has been raised that it may play a central role in the origins of atherosclerosis and heart attacks." (p. 3) The author is director of Pediatrics at Johns Hopkins University School of Medicine. See related articles at [www.fortunecity.com/littleitaly/calabria/138/nutrition/vegetarian/disadvantagemilk.html] and [www.perlhealth.com/chap_7.htm]. This is a rather convincing work, and by an author who cannot be easily dismissed as "a quack."

615. Page, Jake. *Zoo: The Modern Ark.* New York: Facts on File, 190 pp., 1990. (081602345x) How zoos are trading animals so as to reduce genetic inbreeding of unwanted traits ("The Species Survival Plan").

616. Page, Jake. *Smithsonian's New Zoo.* Washington, DC: Smithsonian Institution Press, 208 pp., 1990. (0874747333) "At the zoo, as in any museum, much more goes on behind the scenes than the public sees, and this book shows the full variety of activities at the nation's zoo." (p. 13) Lots of photographs, and useful short essays on research, education, and care-taking at the National Zoo. Most zoos, in spite of their publicly proclaimed high ideals, however, use shady deals to dispose of unwanted surplus animals on to the "canned hunt" and exotic pet markets. See Alan Green's *Animal Underworld* for more details on under-handed zoo practices in chapter 3 of this book.

617. Palika, Liz. *Love on a Leash: Giving Joy to Others through Pet Therapy.* Boulder, CO: Alpine, 162 pp., 1996. (0931866766) See a dog therapy site list at [www.golden-retriever.com/therapy.html].

618. Phillips, David, and Nash, Hugh, eds. *The Condor Question: Captive or Forever-Free.* San Francisco: Friends of the Earth, 297 pp., 1981.

(0913890480) See a biography of Phillips at [www.earthtrust.org/ffbios.html]. I have not read this book, but the title seems to imply that condor captivity is unethical. At the time it was being written, the California condor was nearly extinct, their population decimated by the insecticide DDT. By capturing the few remaining wild condors, scientists were able to save the species and create a breeding population. I do not know many people who would condemn the scientists for their action in the late 1970s. However, there are a few who say that animals should never be captive for any reason, even to save them.

619. Pryor, Karen, and Lorenz, Konrad. *Lads Before the Wind: Diary of a Dolphin Trainer*, 3rd reprint ed. Waltham: Sunshine Books, 278 pp., 1994 (originally published in 1975). (0962401730) Pryor trained dolphins in aquariums in the 1960s and 70s, and tells remarkable stories about dolphin abilities, including creativity. The Foreword by Konrad Lorenz is also notable. Read a brief biography of Pryor at [www.pbs.org/wgbh/nova/vets/biopryor.html]. See links to animal training sites at [www.geocities.com/Athens/Academy/8636/Links2.html].

620. Ritchie, Carson I. A. *Insects, the Creeping Conquerors and Human History*. New York: Elsevier, 139 pp., 1979. (0840766068) "Discusses the effects insects have had on history, art, and medicine." Includes the use of insects for production of silk. See a major insect page with a great bibliography at [www.insects.org/home/index.html]. Read an article opposing the human use of bees and silkworms at [arrs.envirolink.org/ar-voices/silkworms.html]. Insects are not often mentioned in animal rights discussions. They are so generally hated by humans that to seek a discussion of rights for them seems ridiculous. Also, there tends to be little support for a belief that insects may be sentient. Insects, more than any other living and moving creatures, seem to be most like little robots, acting upon "instinct." Is this why we exclude them from debate, or is it just because they are so repulsive?

621. Roberts, Yvonne. *Animal Heroes*. London: Pelham Books, 88 pp., 1990. (0720719372) True animal stories about guide dogs, helping monkeys, brave war animals, hero pets, sheep dogs, police dogs, and so on. There are many books of this type, full of anecdotes about animal deeds. Scientists too often ignore anecdotal evidence because it is not seen by "qualified researchers" in a controlled environment. On the other hand, some folks may read too much into animal behavior. I am reminded of a recent television cartoon where the wife says her dogs are so smart they seem to understand what she says to them. The husband speaks sweetly to the dogs "doggies, wanna go to the pound and be put to sleep now?" as they bounce

happily up and down. Obviously the animals in this case are reacting to the tone of voice, not the words of the owner.

622. Rollin, Bernard E. *An Introduction to Veterinary Medical Ethics: Theory and Cases.* Ames: Iowa State University, 417 pp., 1999. (0813816599) A wonderful and practical book for veterinarians, presented largely in question-and-answer format. There are 82 cases. What should I do if I suspect a pet owner is committing cruelty? What if an owner asks me to euthanize a healthy animal simply for convenience? See a related article at [www.caliban.org/Jo/ast.htm].

623. Rollin, Bernard E. *Farm Animal Welfare: Social, Bioethical and Research Issues.* Ames: Iowa State University, 180 pp., 1995. (0813825636) Includes chapters on beef, swine, dairy, veal, and poultry industries. Rollin may be the most "middle-ground" of the animal rights philosophers, attracting respectful attention from both animal-rights activists and industry representatives. See an extensive article on dairy farming at [www.nal.usda.gov/awic/newsletters/v9n1/9n1arave.htm]. Find many links to animals in ethics questions at [www.ethics.ubc.ca/resources/animal/].

624. Rosenthal, Richard. *K-9 Cops: Stories from America's K-9 Police Units.* New York: Pocket, 360 pp., 1997. (0671000233) See one police K-9 unit homepage at [springfield.missouri.org/gov/spd/K-9s.htm]. A good related links page at [k9cop.com/www.htm].

625. Sainsbury, David. *Farm Animal Welfare: Cattle, Pigs and Poultry.* San Francisco: Harper, 186 pp., 1987. (0003831574) See articles on the subject at [agecon.lib.umn.edu/mn/p91-01.html] and [asci.uvm.edu/bramley/welfare.html].

626. Salisbury, Joyce E. *The Beast Within: Animals in the Middle Ages.* New York: Routledge, 238 pp., 1994. (1415907683) Animals used as food, property, sex objects, beasts of burden. The impact of Christianity on animal lives. See a substantial bibliography of medieval bestiaries (art portraying animals) at [www.clues.abdn.ac.uk:8080/besttest/alt/comment/biblio.html].

627. Sanders, Clinton R. *Understanding Dogs: Living and Working with Canine Companions.* Animals, Culture and Society series. Philadelphia: Temple University, 224 pp., 1999. (1566396891) What dogs mean to us, and what we mean to them. The author is a sociologist, studying human-animal interactions. See articles by the author at [www.psyeta.org/sa/sa2.1/sanders.html] and [www.psyeta.org/sa/sa4.2/sanders2.html].

628. Sawicki, Stephen. *Animal Hospital.* Chicago: Chicago Review Press, 227 pp., 1996. (1556522746) The author spent one year working at the Angell Memorial Animal Hospital in Boston. Reviewed at [www.ipgbook.com/press/anho1.htm].

629. Seguin, Marilyn, and Caso, Adolpho. *Dogs of War: And Stories of Other Beasts of Battle in the Civil War.* Boston: Branden, 155 pp., 1998. (08283-20314) This is an interesting book with short tales about the role of animals in the U.S. Civil War. See a site about war animals that won medals at [homepages. ihug.co.nz/~phil/morean.htm]. Read a substantial article on animals killed during war at [www.geocities.com/RainForest/Canopy/2153/war.htm].

630. Serpell, James A. *In the Company of Animals: A Study of Human-Animal Relationships,* revised ed. Oxford: Basil Blackwell, 215 pp., 1996 (originally published in 1986). (0521577799) The author works in veterinary medicine at the University of Pennsylvania. He wonders about man's relationship to animals and the world, puzzling over the paradox of pet love versus domestic animal hatred. What purpose do animals serve in our society? See an article by the author at [www.psyeta.org/sa/sa3.2/serpell.html].

631. Sheperdson, David J., et al. *Second Nature: Environmental Enrichment for Captive Animals.* Zoo and Aquarium Biology and Conservation series. Washington, DC: Smithsonian, 336 pp., 1998. (1560987456) How creation of proper environments for captive animals leads to more normal behaviors and breeding patterns.

632. Soave, Orland A., and Popejoy, Lori L. *Animal/Human Bond: A Cultural Survey.* Bethesda, MD: International Scholars, 224 pp., 1997. (15729-21048) Discussions of animals as pets, the process of domestication, primate relationships, and so on. See a large paper on the subject at [www.soc.hawaii.edu/leonj/leonj/student3/calvinc/499/pets.html].

633. Stern, Jane, and Stern, Michael. *Dog Eat Dog: a Very Human Book about Dogs and Dog Shows.* New York: Scribners, 192 pp., 1997. (0684822539) The people and rituals of pedigree dog shows. This book is "for anyone who has ever owned a dog, showed a dog, or wondered what could possibly motivate those who do." Read the American Kennel Club rules for shows at [www.akc.org/regs.htm].

634. Stevenson, Peter. *A Far Cry from Noah.* London: Green Print, 1994. (1854250892) An expose about the business of live animal exporting to Europe. The author works for Compassion in World Farming. This is an

indictment of the Ministry for Agriculture. See an article about protests in Britain against live animal exports at [www.spunk.org/texts/actions/sp00 1565.txt]

635. St. John, Patricia. *The Secret Language of Dolphins*. New York: Summit Books, 268 pp., 1991. (0671709798) A very balanced analysis of the issue of keeping dolphins in captivity. The author uses dolphins to try to facilitate communication with autistic children. Read articles on this subject at [fp.premier1.net/iamdavid/healing.html] and [www.thepeacefamily.force9.co.uk/dolphins.html]. See an article against keeping dolphins in captivity at [www.fund.org/facts/dolphins.html]. There are several articles about the dolphin and whale captivity debate at [www.pbs.org/wgbh/pages/frontline/shows/whales/debate].

636. Stuck, Hudson. *Ten Thousand Miles with a Dog Sled*, reprint ed. Battle Creek, MI: Wolfe Publishing, 420 pp., 1988 (originally published in 1914). (0803241925) The adventures and dangers of dog sledding for immense distances. See a list of sled dog books at [www.dogbooks.com/sled.htm] and an article about sled racing at [www.canismajor.com/dog/iditarod.html].

637. Tors, Ivan. *My Life in the Wild: Capturing and Training Animals for Movies*. Boston: Houghton Mifflin, 209 pp., 1979. (0395277663) Read articles against the use of animals in entertainment industries at [arrs.envirolink.org/ar-voices/sea-world.html] and [www.avar.org/pos_stat.Htm #motionpicture].

638. Townsend, Christine. *Pulling the Wool*. Sydney, Australia: Hale and Ironmonger, 1985. A book about the Australian wool and sheep industry. Read a manual on the raising of sheep in Australia at [cyberchino.com.au/SHEEPIN.html] and an anti wool site at PeTA [www.peta-online.org/cmp/ccprodfs7.html]. The case against wool is not an easy one to make. Sheep do not seem to be harmed physically by shearing; in fact, it would be cumbersome to wear that wool in hot weather. There are other considerations, though. Have we altered the sheep genetically to make them carry this unnaturally heavy wool? Do we slaughter local predators in order to keep our valuable wool-bearers safe?

639. Tudge, Colin. *Last Animals at the Zoo: How Mass Extinction Can Be Stopped*. Washington, DC: Island, 266 pp., 1992. (1559631589) The author writes that the four functions of a modern zoo are conservation, education, research, and entertainment. Many of the animal rights arguments in opposition to zoos are related to the fact that zoos keep animals for entertainment purposes; they make money off of visitors. On

the other hand, zoos use much of this money to fund the research, education, and conservation projects that most animal rights activists would support. It is almost like the argument for gambling in some states: if gambling can reduce tax burdens, why not? Well, answer some, it may be immoral. So we are back to the same question: Is it immoral to keep an animal in captivity, even if the captivity is supposed to help the species? See an article about zoo efforts in conservation and research at [www.zod.wau.nl/~www-vh/etho/ISAE-ZOO.html].

640. Unknown. *Animal Husbandry: Directory of Authors of New Medical and Scientific Reviews with Subject Index*. Washington, DC: ABBE, 1996. (0788310305) See a brief article about ancient animal domestication and husbandry at [www.le.ac.uk/archaeology/rug/AR210/TransitionsToFarming/animals.html].

641. Van Etten, Rick. *Tales of Teams: Heartwarming Memories of Hard-working Horses and Mules*. Greendale, WI: Reiman Publications, 162 pp., 1995. (0898211506) See a list of books on work horses at [www.draftresource.com/books/Books.html].

642. Wheale, Peter, and McNally, Ruth, eds. *Animal Genetic Engineering: Of Pigs, Oncomice and Men*. Petersfield, England: Pluto Press, 293 pp., 1995. (074530754x) Twenty contributors. "[W]hat is animal genetic engineering and what place in society should the genetic engineering of animals have?" Read an article about biotechnology at [www.gn.apc.org/eco/resguide/2_18vii.html]. This may be the most important issue of the next century and millenium. Genetic engineering has the potential to completely alter human and animal lives in fundamental ways, at the cellular level, even from birth. There is astounding possible good that might come of it, and terrifying dangers also. Will we proceed with such experiments even before we have decided if genetic engineering is right?

643. White, Joseph J., and Luther, Luana, eds. *Ebony and White: The Story of the Canine Corps*. Wilsonville, OR: Doral, 168 pp., 1996. (094487536x) Dogs in warfare from World War I through Vietnam. The author is now president of the National War Dogs Memorial Fund. See related sites [www.qmfound.com/K-9.htm] and [www.qmmuseum.lee.army.mil/dogs_and_national_defense.htm].

644. Whittemore, Hank, and Hebard, Caroline. *So that Others May Live: Caroline Hebard & Her Search-and-Rescue Dogs*. New York: Bantam, 295 pp., 1995. (0553099515) Hebard is a co founder of the U.S. Disaster Response Team. These are stories of worldwide tragedies (earthquakes, floods, etc.) in which dogs played a role in helping the victims. Find

articles about search and rescue dogs at [members.bellatlantic.net/~k9sar/] and [www.canismajor.com/dog/sandresc.html].

645. Wilson, Cindy C., and Turner, Dennis C., eds. *Companion Animals in Human Health*. Thousand Oaks, CA: Sage, 310 pp., 1998. (0761910611) "To date, studies have shown animal contact could be healthy, contribute to child development of nurturance and self-concept, promote dialogue among family members, children, people with disabilities, and lonely people, contribute to physiological well-being and improvement of select cardiovascular markers and reduce anxiety levels." These were all peer-reviewed papers presented at the 7th Annual Conference on Animals, Health, and Quality of Life. Read a short review at [www.mja.com.au/public/bookroom/chapman/chapman.html].

646. Zayan, Rene, and Dantzer, R., eds. *Social Stress in Domestic Animals: A Seminar in the Community Programme for the Coordination of Agricultural Research*. Current Topics in Veterinary Medicine and Animal Science series. Dordrecht, the Netherlands: Kluwer, 1990. (0792306155) About the reactions of animals to the stresses of overcrowding and transport. Read an article on the transportation of horses at [www.fund.org/facts/horstran. html].

Chapter 5

Animal Populations

The animal rights topic most closely related to the modern environmental movement is that which I have labeled "Animal Populations." The multi-faceted subjects of endangered and extinct species underlie most of the issues in this chapter.

Habitat loss is the single most important factor in the destruction of wild species. Without a place to live, a species cannot survive. Animals require a steady supply of food and water, and territory in which to find those supplies. Animals have become endangered and extinct throughout Earth's history, even without human intervention, but it is human populations that are now causing most of the population problems among animal species. These habitat encroachments come from "urban sprawl," as cities expand to form suburbs; from slash-and-burn farming, which is practiced in many third world nations for quick profit; from timber cutting; from pollution and oil spills; and from recreationists and parks. Not all of these actions are done intentionally to cause harm, but it would be naïve to believe that greed and negligence do not play some role.

Habitat restoration has become a hot-button topic, as developers and land owners fear that the government is usurping their land. In some ways they are correct, particularly in the lands west of the Mississippi River, where the federal government owns and controls a vast majority of the property. Furthermore, is it not hypocritical of many environmental and animal-rights organizations to demand land restrictions upon lands that are far away and over which they have no claim? How loudly would they be crying for governmental regulation if the land belonged to them? No one likes to have their property regulated out of usefulness, where homes or roads cannot be built. On the other hand, if regulations are not made, what will protect the species? One problem in this discussion is that "species" itself is a poorly defined term, argued over by scientists constantly. What constitutes a species? Is a woodpecker with an orange head a different species than the one with the red head? Is the wasp with two abdomen stripes a different species than its single-striped cousin? In times past a species was defined more specifically, regarding reproductive abilities in like types. But the nature of our current Endangered Species Act leaves the question primarily a legal one, for courts, not scientists, to decide. Thus, species protection has become a political football, going back and forth without an ultimate victor, thus far. If you wish to protect a species, you have to sue in court so that judges must make injunctions against development. It is quite a complicated process.

Who is to blame for endangered species? Is it the developers who level the land, or the buyers who demand homes outside of the city? Or is it something else? Socialists blame it on capitalism. Capitalists blame it on a few unscrupulous individuals. The unscrupulous individuals blame it on the buyers.

There is, of course, a problem with overpopulation of some animals. Not all species are endangered. You will frequently hear of major crises in places like Australia, where rabbit or kangaroo or mouse populations explode because the predators have all been killed. This is a problem in the United States also. Predator animals kill game animals, controlling their populations. But people are afraid of predators and kill them. I recently read a newspaper article by an Ohio woman, saying that she heard there was a cougar somewhere in eastern Ohio, and it must be instantly hunted down and shot because it might harm her grandchildren. Is it lack of information and education that leads to this sort of irrational fear (there is one cougar somewhere in the state, so we must hunt it down and kill it)? The states used to (and a few still do) pay bounties for predator hides: people who bring in a wolf or cougar pelt get some money. Once the predators were mostly killed off (now surviving mainly in zoos), the game species had no one to control their populations. In the United States we have been solving this problem by having human hunters "cull" the herds. Predators have been killed in many ways, through trapping, hunting, and poison. Remarkably, predators are killed by the U.S. government at taxpayer expense on federal lands; because the government leases the land to ranchers, whenever a sheep or calf is killed by wolves or cougars, out come the helicopters and snipers to hunt them down. Lest you immediately jump to the conclusion that the government should stop leasing land to ranchers, however, consider that the United States government directly owns or controls some sixty to seventy percent of the land west of the Mississippi River. So where else will ranchers go to graze their cattle? The vegetarian solution would be to stop raising cattle; but vegetarianism is currently a minority preference.

Overpopulation is a problem. When there are too many members of one species in a given area, they will overeat their food supply. In short, they will starve themselves to death, and possibly starve other species also, that share that food supply. What else can happen? Overgrazing of cattle or wild species can lead to the same problems that over plowing can; the plants are no longer available to hold down the topsoil, creating a dust bowl, with wind and drought eroding and blowing away the soil.

The controversy will not be dying down in the near future. The few recent attempts at predator reintroduction, of the lynx (a small version of a mountain lion) to Colorado and the wolf to Yellowstone National Park, have led to major confrontations with locals. Mothers fear for their children; ranchers fear for their livestock. There are probably more fatal hunting accidents in a single state than there are humans killed by wild predators in the whole United States in any given year.

Predators are not the only controversial wild land-using animal in the United States. Ranchers want to remove the wild horses from the western states, where they compete with cows for the grass. This is similar to the situation in Africa, where growing human populations are upset by the elephants and wildebeest who sneak onto the farms at night to graze. Can you wild animals all be kept inside the park? When they step outside, they may be shot by angry farmers. It is hard to blame the farmers; they seek to make a living, and an elephant herd can surely do a lot of damage. On the other hand, where else are the elephants going to find food when African humans are filling the land? Now elephants and other species are being "culled," having their populations intentionally reduced to minimize these sorts of problems. However, the effectiveness and long-term effects of culling are not yet known. Is the smaller population a viable and healthy one? Are the authorities killing the older ones, and leaving the younger ones without experienced guardians? And how effective has the ivory ban been? Some say that it is like the drug war. By making drugs illegal, they have become more expensive, and thus more worthwhile to risk obtaining and selling. Is poaching for ivory reduced now that ivory is illegal? There are several books in this chapter that discuss this question.

Whale populations dwindled in the late twentieth century under the pressure of Japanese, Russian, and northern European mechanized whaling operations. Some species, like the grey whale, appear to be recovering since the whaling reductions of the late 1970s, spearheaded by Greenpeace.

Critics say that some animal-rights efforts to save populations are having the opposite effect. Janice Scott Henke (see chapter 3) says that seals are becoming overpopulated and ruining the northern fishing industry, due to the anti-seal-pelt campaigns. They also say that such campaigns hurt indigenous and native peoples the most, like the Inuit, who live on their seal and narwhal catches. Is this in fact "cultural imperialism?" Are the liberals of first-world nations, who live comfortably already, demanding that other poorer peoples save their animals at their expense? How will they survive (let alone prosper) if seals are their only means of livelihood? Should they just move out?

One of the major philosophical differences between the opposing camps is the distinction between conservation and preservation. Animal-rights people and environmentalists tend to believe in "preservation," which means the complete protection of habitat lands from any human use or intrusion. In recent years, the Clinton/Gore administration has forwarded this cause by the banning of motorized water craft, like jet skis and motor boats, from most national parks. Officials are also destroying roadways, and cutting back on air space over parks so that the animals and shrubs are not bothered by air and noise pollution. These are protectionist tactics. On the other side are the conservationists, who tend to believe in "sustainable use," which means that people must be allowed to use some of the resources, but not all. They must derive some economic gain from land protection for it to be effective and fair. Conservationists say that hunting is good in moderation, logging is good in moderation, and so on.

The question of "ecotourism" is somewhat more complex than it might seem. On the surface, it is a great idea, encouraging local peoples to preserve their wildlife and scenic areas so that they will profit from the visiting tourists. In theory it sounds promising, but in reality things do not always work out as foreseen. For one thing, the economic system of the country has a major impact on the practices. In many political systems, the only people who see any money from the ventures are those who could buy into the project in the first place. The locals never see any of the benefits; the money flows away to a few owners or politicians. Furthermore, some habitats become overwhelmed by the massive influx of tourists. When the numbers of people, or types of recreation allowed, are not supervised, all kinds of damage may be done. For example, at many popular underground caverns, people have broken off the stalactite and stalagmite tips to take home as souvenirs. After a few years, much of the beauty is lessened by irresponsible visitors. There is some work to be done on ecotourism.

Who is to make the laws regarding species? Right now, the United Nations and such organizations, like the World Court, have little power. About the only tools they have are economic sanctions, and economic sanctions are not even used often for human rights violations, let alone animal rights or environmental violations. Do we really want a world organization making decisions about "our" nation's land? Do we even want the federal government making such decisions? How about our states, or is this solely a local matter? To go completely into the *laissez-faire* philosophy, maybe it is no one's business but the owners. This becomes especially complex with species who travel between different regions, like birds, whales, or turtles. If poor Mexicans continue to kill the female turtles as they climb the beach to lay eggs, to sell the shells and make soup, several species of sea turtles will be extinct in the next decade. So does the United States have the right to demand enforcement of the laws against killing endangered turtles in Mexico?

Another problem of a different sort is bioinvasion. Bioinvasion is the undesirable introduction of a strange species to a new territory, where the species has no natural predators and may overwhelm the local ecosystem. Some recent examples of this are the zebra mussels, which have nearly clogged up Great

Lakes waterways with layers of mollusks, and a certain form of Asian termite that is spreading through Louisiana, destroying old wooden structures. This type of situation is the kind that creates odd questions for animal rights people. In this case, we must actually seek to control or even wipe out the invading species, in order to protect the native species that cannot compete!

Animal population questions are no easier than those in the earlier chapters. All of these questions have direct bearing and effects on human lives, good and bad.

---

647. Ackerman, Diane. *The Rarest of the Rare: Vanishing Animals, Timeless Worlds*, reprint ed. New York: Vintage, 208 pp., 1997 (originally published in 1995). (0679776230) Poetic observations on Ackerman's travels to see endangered habitats. She says that every extinction we cause may be putting mankind at risk. Read a review at [www.thebookery.com/Book store/details.cfm/Authors/132.htm]. See many biodiversity links at [www. heritage.tnc.org/oth svrs.html].

648. Adams, Jonathan S., and McShane, Thomas O. *The Myth of Wild Africa: Conservation without Illusion*, reprint ed. Berkeley: University of California, 282 pp., 1996 (originally published in 1992). (0520206711) Attacks the commonly held belief that Africans do not care about their wildlife, saying that "celebrity scientists" like Dian Fossey always blame the locals. Preservation efforts cannot ignore the needs of the local human populations. Read the publisher review at [www.ucpress.edu/books/pages/6901. html]. While it is certainly true that blame does not rest entirely on the local indigenous human populations, it would be reckless not to focus on their involvement. It is much easier on a geographic level to stop poaching at the site of the act than to stop foreigners worldwide from demanding the animals' products. Sure, the foreigners have a lot of the blame, but the actual killing and removal occurs at the local level and must be stopped there if a species is to survive. The larger question of improving the economies of the poor indigenous peoples, to reduce the desire for poaching and quick profits, should be addressed also, as the author rightly posits.

649. Adams, Lowell, Adams, W., and Geis, A. D. *Effects of Highways on Wildlife Populations*. Columbia, MD: Department of Transportation, 142 pp., 1981. See a major article on this subject at [www.flora.org/afo/afz/ issue10.html]. I would guess that highway "road kill" of animals takes a much higher toll than legal hunting. You cannot take a drive without seeing some creature's body lying along side the road, the victim of fast moving tires. Some very simple improvements can be made to toll roads to

reduce the kill, leaving occasional breaks in the concrete barriers so that the animals can find a way across without being trapped at the median. I have also seen little whistles mounted on bumpers, which supposedly emit a sound to frighten the animals away from the roads, but I do not know if they are effective. Read a summary of this report at [www2.ncsu.edu/unity/lockers/project/openspace/biblio/adams.htm].

650. Adler, Bill, Jr. *Outwitting Critters: A Surefire Manual for Confronting Devious Animals and Winning*, reprint ed. New York: Lyons & Burford, 256 pp., 1997 (originally published in 1992). (1558215239) "[T]he definitive book on coping with the nettlesome side of nature." (backcover) Chapters are organized by species. See an interesting store that sells various animal-control devices at [www.deerbusters.com/].

651. Allen, Durward Leon. *The Wolves of Minong: Their Vital Role in a Wild Community*, reprint ed. Ann Arbor Paperback series. Lansing: University of Michigan, 499 pp., 1993 (originally published in 1979). (047208237x) How wolves and moose populations survive together in Michigan. Read an obituary and biography of the author at [www.wolf.org/GH/Preview/spring98D5c.html]. See a bibliography of wolves at [www.lochnet.net/wolfen/animalbooks/wolfbook.htm] and an article about their social lives at [www.fcinet.com/yelpaper/this_issue/about_wolves.html]. Wolves have been greatly misunderstood, blamed for all sorts of wildlife carnage that they probably have nothing to do with. As Farley Mowat found in *Never Cry Wolf*, they prey on the old and weak members of ungulate herds, and eat mice and rabbits. They do not decimate elk and caribou populations. Predators are sadly missed in the United States, leading to overpopulation of prey species.

652. Allen, Kenneth Radway. *The Conservation and Management of Whales*. Seattle: University of Washington, 107 pp., 1980. (0295957069) See a related bibliography and links at [www.whalesci.org/whalebooks.html], and an article at [www.highnorth.no/th-in-wh.htm]. The protection of whales and other marine mammals is greatly complicated by their migration patterns, which take them into the waters of many nations that do not enforce protection efforts.

653. Amory, Cleveland. *Man Kind? Our Incredible War on Wildlife*. New York: Harper & Row, 372 pp., 1974. (0060100923) The author uses critical irony in chapter headings, such as "Support Your Right to Arm Bears" and "Real People Wear Fake Furs." He founded The Fund for Animals. He became an animal rights activist after witnessing a bullfight in Mexico. He says that hunters should not control wildlife agencies. Read a biography and obituary at [www.animalnews.com/news/nytobit.htm].

654. Anchel, Marjorie, ed. *Overpopulation of Cats and Dogs: Causes, Effects, and Prevention.* Conference Proceedings, New York State Humane Association, 1987. New York: Fordham University, 260 pp., 1990. (08232-12963) Papers (speeches) by Ingrid Newkirk, Tom Regan, and others. See related articles at [www.canismajor.com/dog/overpop.html] and [www.spayusa.org/overpopulation.html]. Some anti-animal-rights researchers claim that animal-rights people and animal shelters greatly exaggerate the number of creatures euthanized in shelters each year, since it increases revenue donations and keeps the idea newsworthy. I do not know who is telling the truth. At any rate, whether exaggerated or not, a lot of animals are euthanized every year.

655. Andrewartha, Herbert G., and Birch, L. C. *The Ecological Web: More on the Distribution and Abundance of Animals*, updated ed. Chicago: University of Chicago, 506 pp., 1986 (originally published in 1954). (02260-20347) Detailed case histories and fieldwork. See the author biography at [macfarlane.asap.unimelb.edu.au/bsparcs/aasmemoirs/andrewar.htm].

656. Baker, Ron. *The American Hunting Myth.* New York: Vantage, 287 pp., 1985. (0533063442) How hunter-dominated state and federal wildlife agencies are systematically destroying America's wildlife. A lot of information about deer and their supposed "overpopulation" which agencies claim to be fighting. Read an article by the author at [arrs.envirolink.org/ar-voices/hunting_extinction.html]. Read an attack on the surplus population argument at [arrs.envirolink.org/ar-voices/hunt.html]. See related articles at [www.fund.org/facts/overview.html] and [www.api4animals.org/HuntingFactSheet.htm]. Hunters and conservationists often claim that deer hunting is good and helps the species, since the animals are overpopulated (due to lack of natural predators) and would starve during the winter. Some animal rights proponents say that the population levels are kept artificially high by intervention by wildlife agencies. The thesis is partially true, in my experience, that wildlife agencies are heavily influenced by hunters. However, these hunters are not destroying our wildlife. They perceive themselves to be taking "surplus" animals, which no longer have natural predators to reduce their population. Hunters are not allowed to destroy a species, legally, and the poachers certainly do not control the wildlife agencies. Perhaps hunting is unethical, but legal hunting in America is not destroying any species. The species being hunted are those that are quite common. It is not in the hunters' wishes to destroy a species; the hunting would end, in such a case. Most hunters are conservationists, wanting a long-term supply of game for sport.

657. Balouet, Jean-Christophe, et al. *Extinct Species of the World: 40,000 Years of Conflict.* New York: Barron's, 192 pp., 1990. (0812057996) More than

300 species considered. See a related bibliography at [www.heritage research.com/wildbib.htm].

658. Barker, Rocky. *Saving All the Parts: Reconciling Economics and the Endangered Species Act.* Washington, DC: Island, 272 pp., 1993. (15596-32011) This work focuses on owls, salmon, wolves, and bears in the Pacific Northwest of the United States. The author is a journalist from Idaho. Read an article by the author at [www.webcom.com/gallatin/pub/GI.RW96. Barker1.html]. See a bibliography at [www.calacademy.org/research/ library.biodiv/biblio/endanger.htm].

659. Barthlott, W., and Winiger, M., eds. *Biodiversity: A Challenge for Development Research and Policy.* New York: Springer-Verlag, 464 pp., 1998. (3540639497) "This volume examines the concerted actions taken by governments . . . and the scientific community for the preservation of biodiversity." See a substantial paper about the efforts of indigenous peoples to sustain biodiversity in their lands at [www.ecouncil.ac.cr/ rio/focus/report/english/ipbn.htm].

660. Baskin, Yvonne. *The Work of Nature: How the Diversity of Life Sustains Us.* Washington, DC: Island, 288 pp., 1998. (1559635193) How the loss of biodiversity may present unforeseen threats to human life. The foreword is written by Paul Ehrlich. Reviewed at [www.sdearthtimes.com/et0797/ et0797s14.html].

661. Beacham, Walton, et al., eds. *World Wildlife Fund Guide to Extinct Species of Modern Times.* Osprey, FL: Beacham, 410 pp., 1997. (09338-33407) This is a multiple volume set. Find a good bibliography and links at [www.naturalenviron.com/endangeredsp.html].

662. Bean, Michael J., and Rowland, Melanie J. *The Evolution of National Wildlife Law*, 3rd revised and expanded ed. New York: Praeger, 544 pp., 1998 (originally published in 1977). (0275959899) "[T]he standard reference, the essential companion for anyone seeking to understand the numerous and complex statutes, regulations, and court decisions on wildlife law." (backcover) Scholarly and objective. See a brief biography of Michael Bean at [www.edf.org/pubs/ExpertGuide/bean_michael.html].

663. Beatley, Timothy. *Habitat Conservation Planning: Endangered Species and Urban Growth.* Austin: University of Texas, 272 pp., 1994. (02927-08068) See an article at [www.ncedr.org/research/hcp.htm] and an article with bibliography on ecological planning at [www.gn.apc.org/eco/res guide/2_6.html].

664. Begon, Michael, et al. *Population Ecology: A Unified Study of Animals and Plants*, 3rd ed. Cambridge: Blackwell Science, 256 pp., 1996. (0632034785) Widely used as a textbook. See a Smithsonian Institution bibliography of animal populations at [www.si.edu/resource/faq/nmnh/endsp1.htm].

665. Belanger, Dian Olson. *Managing American Wildlife: A History of the International Association of Fish and Wildlife Agencies*. Amherst: University of Massachusetts, 264 pp., 1988. (0870236083) The work includes a prominent discussion on the debate over state versus federal jurisdiction over wildlife. Jurisdiction questions between state and federal agencies have been standard in the United States for many decades. Generally, since the Civil War, federal powers have been growing at the expense of state powers. We have state parks, national parks, and local parks. The question of who enforces wildlife laws in each of these locales is a complex one. The author is a vice president of the American Association of University Women.

666. Benton, Ted. *Natural Relations: Ecology, Animal Rights, and Social Justice*. London: Verso, 280 pp., 1993. (0860913937) An attempt to reconcile animal rights with socialism and environmentalism, which have usually been at odds. Karl Marx's views are discussed in detail. Some reviewers found the structure and language to be difficult. See a related periodical at [www.guilford.com/periodicals/jncn.htm].

667. Blair, Cornelia B., et al., ed. *Endangered Species — Must They Disappear?* Information Series on Current Topics. Wylie, TX: Information Plus, 144 pp., 1996. (1573020249) See a report on the effectiveness (or lack thereof) of the Endangered Species Act at [www.edf.org/pubs/Reports/help-esa/].

668. Bolton, Melvin, ed. *Conservation and the Use of Wildlife Resources*. Conservation Biology series #8. London: Chapman & Hall, 278 pp., 1997. (0412713500) A calm examination of the sustainable use theory, and some discussion of ecotourism. Case studies include the standard analyses of primate and bird populations, but also the unusual addition of clam populations. Read one view on the sustainable use theory at [arrs.envirolink.org/ar-voices/wise_use.html]. "Conservation" and "preservation" are two distinctly different words in animal rights or environmental discussions. Conservation tends to be the more moderate position, wherein resources are "utilized" in a "wise" manner. Preservation is usually the more liberal position, wherein resources should not be used much at all, but protected from human use. Hunters tend to call themselves

conservationists, while most environmentalists and animal rights supporters would often call themselves preservationists.

669.Bonner, Raymond. *At the Hand of Man: Peril and Hope for Africa's Wildlife*. New York: Vintage, 322 pp., 1994. (0679400087) A controversial and important book about African wildlife in relation to economic realities of the indigenous peoples. How wildlife organizations sometimes suppress and ignore truth to frame the debate in a simple way (for donors). The author shows hope in the idea of "sustainable utilization," which includes the culling of local herds. He says this is more effective than Western nations demanded total bans on killing. Bonner calls animal-rights groups "ecocolonialists," and says that the ivory ban is merely public relations noise. Read a review at [www.naiaonline.org/book9510.html]. Find a bibliography on African wildlife at [www.awr.net/read.html].

670.Boo, Elizabeth. *Ecotourism: The Potentials and Pitfalls: Country Case Studies*. Washington, DC: World Wildlife Fund, 288 pp., 1992. (0942635159) Two volumes. See a bibliography at [www.eduweb.com/schaller/RBbibliography.html], a site with articles and a bookstore at [www.ecotourism.org], and a substantial article at [www.ecotourism.org/textfiles/mccool.txt]. Ecotourism is an innovative idea to give local human populations a way of making money by *not* killing the animals and destroying the environment. In theory it sounds very good. In practice though, there are many problems. Human tourists by nature wreak much havoc, unintentionally, such as feeding the animals with junk food at the zoo. The animals become tame and accustomed to visitors, more reliant on handouts than wild animals. Roads are built; vehicles and campers pollute; noise frightens smaller skittish species. Furthermore, in many cases the ecotourist money all goes to a few locals, and most of the local people reap no benefits from the practice. Maybe ecotourism will work, but there are some obstacles to overcome in the attempt.

671.Bowles, Marlin L., and Whelan, Christopher J. *Restoration of Endangered Species: Conceptual Issues, Planning, and Implementation*. Cambridge, MA: Cambridge University, 394 pp., 1994. (0521418631) Wildlife populations are dwindling as human impact on habitat increases. Examines the recovery of some rare species. See a bibliography of ecological restoration at [www.forestry.uga.edu/docs/for97-10.html] and links to wildlife rehabilitation sites at [www.oneeyedcat.com/Healers_of_the_Wild/links.html].

672.Bright, Chris. *Life Out of Bounds: Bioinvasion in a Borderless World*. Worldwatch Environmental Alert series. Washington, DC: Worldwatch, 287 pp., 1998. (0393318141) On the problem of non-indigenous species entering new habitats where there are no natural predators. See a good review at [www.newscientist.com/ns/981107/review.html] and a brief biography of the author at [www.worldwatch.org/bios/bright]. Read an article by the author at [www.worldwatch.org/alerts/pr981010.html]. Read an online book, *Stemming the Tide: Controlling Introductions of Non-indigenous Species by Ships' Ballast Water*, at [books.nap.edu/bo_book/0309055377/html/index.html].

673. Brown, W. *Acid Rain and Wildlife*. Washington, DC: Environmental Defense Fund, 1981. See the site for the U.S. Acid Rain program at [www.epa.gov/acidrain/student/student2.html]. For some years it has been thought that acid rain may be the main reason for the radical decrease in amphibian populations around the world because frogs' sensitive skins are very susceptible to pollutants. Another recent theory, however, says it is a bacterial type of infection. See a great site regarding the amphibian mystery at [www.tigerherbs.com/eclectica/earthcrash/subject/deformities.html].

674. Bryant, Peter J. *Biodiversity and Conservation: A Hypertext Book*. Irvine: University of California, 1999. A substantial online book that you can read at [darwin.bio.uci.edu/~sustain/bio65.Titlpage.htm#Tableof Contents].

675. Buchanan, Carol. *The Wildlife Sanctuary Garden*. Berkeley, CA: Ten Speed Press, 192 pp., 1999. (1580080022) How to create a haven in your backyard for many wild creatures. There is an e-mail forum for gardeners who want to attract wildlife at [www.nature.net/forums/garden/]. As cities have grown in population, they have spread out and created massive suburbs. As the people move outward into the country, wildlife either adapts (becoming scavengers) or is displaced. These wildlife sanctuary gardens are one way to help some wild creatures, especially birds. See some bird watching links at [birdsource.cornell.edu].

676. Burton, John A., ed. *The Atlas of Endangered Species*, 2nd ed. New York: Macmillan, 256 pp., 1998 (originally published in 1991). (0028650344) The atlas is arranged by climatic regions, and includes some plant species as well as animals. Lovely, but rather expensive. National Geographic has a nice interactive atlas of endangered species at [magma.Nationalgeographic.com/2000/biodiversity/index.cfm].

677. Cadieux, Charles L. *Wildlife Extinction*. Washington, DC: Stone Wall, 259 pp., 1991. (0913276359) This prolific author (former president of the Outdoor Writers Association of America) says that uncontrolled human population growth is hurting all species on Earth. Includes case studies on owls, ferrets, condors, and whales. See a site of Canadian endangered animal species at [www.cws-scf.ec.gc.ca/es/endan_e.html].

678. Carson, Rachel. *Silent Spring*, reprint ed. Boston: Houghton Mifflin, 368 pp., 1994 (originally published in 1962). (0395683297) This marine biologist spent a career with the U.S. Fish and Wildlife Service. This book led President Kennedy to set up a special panel to study the problem of pollution by pesticides, and started a wave of environmental legislation. Many scholars credit this work as the beginning of the American environmental movement. See a web page about the author at [www.rachelcarson.org/]. Read an article and bibliography about pollution at [www.gn.apc.org/eco/resguide/2_8.html].

679. Caughley, G., and Sinclair, A. R. E. *Wildlife Ecology and Management*. Cambridge: Blackwell Science, 344 pp., 1994. (0865421447) See a good bibliography of wildlife ecology materials at [wildlife.wisc.edu/grad/readinglist.htm].

680. Chadwick, Douglas H. *The Company We Keep: America's Endangered Species*. Washington, DC: National Geographic, 157 pp., 1997. (07922-71327) A few essays, but mostly photographs. Read a brief article about endangered species at [www-honors.ucdavis.edu/species/biological].

681. Clark, Tim W. *Averting Extinction: Reconstructing Endangered Species Recovery*. New Haven: Yale University, 288 pp., 1997. (0300068476) Largely a study of the effort to save the black footed ferret in America. See a paper on species reintroduction at [arrs.envirolink.org/ar-voices/reintro.html].

682. Clark, Tim W. *Endangered Species Recovery: Finding the Lessons, Improving the Process*. Washington, DC: Island, 512 pp., 1994. (15596-32712) Based on a conference at the University of Michigan in 1993. Read an article by the author at [www.umich.edu/~esupdate/library/96.09/clark.html]. See a good bibliography of endangered species at [www.lib.umd.edu/UMCP/MCK/GUIDES/endangered_species.html].

683. Cohn, Priscilla N., et al., eds. *Contraception in Wildlife*. Lewiston, NY: Mellen, 358 pp., 1996. (0773488278) "The need to minimize unwanted reproduction in captive populations of wildlife has stimulated research with different methods and drugs." (p. 1) Cohn writes an excellent preface on hunting and culling. Read an article of various methods that states are trying to reduce deer populations at [www.dgif.state.va.us/hunting/eval.htm].

684. Cohn, Priscilla N., ed. *Ethics and Wildlife*. Mellen Animal Rights Library. Lewiston, NY: Mellen, 276 pp., 1999. (0773487123) The author is a philosophy professor at Pennsylvania State University at Abington. Writers

include Evelyn Pluhar, Gary Francione, Tom Regan, and Andrew Linzey. "Even the animal rights movement has not paid much attention to wildlife." (p. iii) See the table of contents at [www.mellenpress.com/html/cohnethi.html]. Sadly, it is true that wildlife usually gets "the short end of the stick" in the animal rights debate. Pets and domestic animals get most of the attention.

685. Collard, Andree, and Contrucci, Joyce. *Rape of the Wild: Man's Violence Against Animals and the Earth*. Bloomington: Indiana University, 187 pp., 1989. (0253205190) "The monies poured into the manufacture of food additives, cosmetics, plastics and endless new chemicals . . . should be channeled into banning all such products." This book is a strong feminist view. Read the table of contents at [www.earthfriendlybooks.com/Sections/catalog/nature/0253205190.html].

686. Conniff, Richard. *Every Creeping Thing: True Tales of Faintly Repulsive Wildlife*. New York: Henry Holt, 256 pp., 1999. (0805056971) Case studies of wildlife as not-so-cuddly-creatures. We should not try to turn the world into a giant petting zoo, which is not respectful to the dangers and mysteries of animals. The author, however, offers no alternatives to the use of zoos to prevent their possible extinctions. See the publisher's advertisement at [hholt.com/98-2hh/everycreepingthing-htm]. It is true that humans naturally extend our sympathy toward the cutest animals, like dolphins and panda bears. Insects and reptiles are not so popular and do not receive the same degree of support and protection against extinction.

687. Deblieu, Jan. *Meant to Be Wild: The Struggle to Save Endangered Species through Captive Breeding*, reprint ed. Golden, CO: Fulcrum, 340 pp., 1993 (originally published in 1991). (1555911668) Specifically about North American animals, the author wonders if we are unintentionally altering their natures in seeking to save them. A link to read an excerpt can be found at [fulcrum-books.com/ html/meant_to_be_wild.html].

688. Decker, Daniel J., ed. *Valuing Wildlife: Economic and Social Perspectives*. Boulder, CO: Westview, 424 pp., 1987. (0813371201) See a list of the author's works at [www.dnr.cornell.edu/hdru/staff/djdpubs.htm]. Read an article on the subject at [bluegoose.arw.r9.fws.gov/Library/EconBene.pdf].

689. DiSilvestro, Roger L. *The Endangered Kingdom: The Struggle to Save America's Wildlife*. Wiley Science editions. New York: Wiley, 241 pp., 1991. (0471606006) What America looked like before humans came, with detailed accounts of 12 species. Surveys our history of destruction and protection, with a study of wildlife management policies. Read an article by

the author which begins at [www.defenders.org/fish01.html]. See a related article at [www.refugenet.com/care.htm].

690.Douglas-Hamilton, Iain, et al. *Battle for the Elephants*. New York: Viking Penguin, 350 pp., 1992. (0670840033) The author and his wife are biologists who have written insightful books about elephants. This work is particularly focused on the hows and whys of the ivory ban of 1989. See an article at [www.africalibrary.org/course/papers/porter&dorrick.html]. Links to elephant sites can be found at [wildheart.com/wwwlinks/main_links.html]. Read the African Elephant Conservation Act at [ipl.unm.edu/cwl/fedbook/afeleph.html].

691.Dunlap, Thomas R. *Saving America's Wildlife: Ecology and the American Mind, 1850-1990*, reprint ed. Princeton, NJ: Princeton University, 222 pp., 1991. (069100613x) Why do we now wish to bring back wolves, when in 1850 they were shot on sight? What cultural changes have occurred? The author is a history professor at Texas A&M University. See a bibliography of American environmental history at [h-net2.msu.edu/~aseh/bibs/floresbib.html].

692.Durrell, Gerald M. *The Overloaded Ark*. Winchester: Faber & Faber, 212 pp., 1987 (originally published in 1953). (0670532819) This author has written dozens of books about wildlife. He was a naturalist who traveled the jungles of Cameroon (and other places) capturing animals to be displayed in British zoos. See biographies of the author at [www.thewildones.org/gerald.html] and [www.geocities.com/Heartland/Plains/2007/durrell.html], and a list of Durrell's books at [www.geocities.com/RainForest/Canopy/6364/durrell_2.html].

693.Easterbrook, Gregg. *A Moment on the Earth: The Coming Age of Environmental Optimism*. New York: Viking, 745 pp., 1995. (01401-54515) Paul Ehrlich says this book "contains so many serious errors that it has spawned a virtual cottage industry among scientists to correct them." Easterbrook may be correct about major improvements that first-world nations have taken regarding the environment (contrary to Ehrlich's views), but he seems to ignore the destruction of third-world nations' environments. The book includes chapters on animal species and spotted owls. Reviewed at [realnews.org/oped/earthrev.htm]. See bibliographies of similar books (counter environmental) at [www.sepp.org.othersources.html] and [www.sepp.org/recread.html].

694.Ehrenfeld, David W., ed. *Genes, Populations, and Species*. Readings from Conservation Biology series. Boston: Blackwell, 1995. (0865424527) See the author's page at [www.rci.rutgers.edu/~deenr/DWE.html].

695. Ehrenfeld, David W., ed. *To Preserve Biodiversity: An Overview*. Readings from Conservation Biology series. Boston: Blackwell, 249 pp., 1995. (0865424519)

696. Ehrenfeld, David W., ed. *Wildlife & Forests*. Readings from Conservation Biology series. Boston: Blackwell, 248 pp., 1995. (086542543) See an article and bibliography about forestry at [www.gn.apc.org/eco/resguide/2_13.html].

697. Ehrlich, Paul R. *The Population Bomb*, reprint ed. New York: Buccanneer Books, 223 pp., 1997 (originally published in 1968). (1568495870) This famous book caused great alarm when released in the late 1960s. Ehrlich predicted widespread famine and death due to overpopulation by 1980, but these alarmist predictions have not yet materialized. Read a negative review at [www.carnell.com/population/paul_ehrlich.html]. See an interview with the author at [www.emagazine.com/november-december_1996/1196conv.html]. See an article discussing human beings as a "cancer" on the Earth at [www.islandone.org/LEOBiblio/SPBI1HC.HTM] and one plan to cut the world's human population back to one billion (down from six billion) at [www.dismantle.org/pop.htm].

698. Ehrlich, Paul R., and Ehrlich, A. *Extinction: The Causes and Consequences of the Disappearance of Species*. New York: Random House, 305 pp., 1981. (0394513126) Read an article on biodiversity by the author at [www.defenders.org/ehrlich01.html]. Read a brief biography at [www.awards.heinz.org/ehrlich.html].

699. Ehrlich, Paul R., et al. *Birds in Jeopardy: The Imperiled and Extinct Birds of the United States and Canada, including Hawaii and Puerto Rico*. Stanford, CA: Stanford University, 259 pp., 1992. (0804719810) See a list of Ehrlich's books at [www.geocities.com/RainForest/3621/BS-EHRL.HTM]. The bookstore at Amazon.com has a good review. See a newspaper article by Ehrlich (and others) on extinction at [victoria.tc.ca/environment/biodiversity/extinction.ecol.html].

700. Eldredge, Niles. *Life in the Balance: Humanity and the Biodiversity Crisis*. Princeton, NJ: Princeton University, 224 pp., 1998. (0691001251) How our fate is bound up with Earth's fate. We are now in the sixth mass extinction. Reviewed at [www.newscientist.com/ns/980620/ review.html].

701. Eldredge, Niles. *The Miner's Canary: Unraveling the Mysteries of Extinction*. Princeton, NJ: Princeton University, 272 pp., 1994. (0691036551) The author is a prolific and famous paleontologist (fossil hunter). The analogy of the miner's canary in relation to extinction is a very potent

one. If the bird died in the mine, it was known that the air was going bad, and soon the human workers would be in danger. Some scientists say that the numerous modern extinctions are a bad sign for human life on Earth. Read an interview with the author at [www.edge.org/documents/Third Culture/m-Ch.6.html].

702. Ellis, B. J. *Paws for Thought: A Look at the Conflicts, Questions & Challenges of Animal Euthanasia*. Columbia, SC: Paw Print Press, 137 pp., 1993. (0963670808) A book to help people understand the reasons for euthanasia and to promote understanding of the work of the people who must perform the procedure on animals. This is a neglected topic. The people at animal shelters who do the "dirty work" probably need a lot of emotional support but instead are reviled as killers by some. Few workers probably enjoy doing this kind of work. See the homepage of the National Animal Control Association, which has links, at [www.netplace.net/naca/]. Find the publisher's brief homepage at [www.yearbooknews.com/ html/PawPrint.html].

703. Facklam, Margery. *Wild Animals, Gentle Women*. New York: Harcourt Brace Jovanovich, 139 pp., 1978. (015296987x) Brief biographies of eleven female scientists who have studied wildlife, including Jane Goodall and Dian Fossey. Includes a chapter called "Is Animal Watching for You?" for potential successors to these sorts of studies.

704. Favre, David S. *Wildlife Law: Cases, Law, and Policy*, 2nd ed. Sterling Heights, MI: Lupus, 444 pp., 1991. (1879581035) Read an article by the author on sport hunting at [www.spiritone.com/~orsierra/rogue/creature/ hunting.htm].

705. Feshbach, Murray, and Friendly, Alfred. *Ecological Disaster: Cleaning Up the Hidden Legacy of the Soviet Regime*. New York: Twentieth Century Fund Press, 1995. Read the introduction at [epn.org/tch/xxfesh03.html]. The author wrote a similar book in 1991 called *Ecocide in the USSR*. Westerners have no idea how badly the environment was damaged in Soviet Russia; communism was not the hero of environmentalism, nor was capitalism the major offender of environmentalism. Some cities, like Chelyabinsk, were completely littered with nuclear waste dumps. The Chernobyl nuclear reactor ruined thousands of square miles of the Ukraine and Belarus. Soviet naval ships simply dumped old reactors over the side into the Arctic Ocean. Some biological weapons were buried in Russian lakes. Russia is in grave ecological danger, more than any Western nation.

706. Fischer, Hank. *Wolf Wars*. Helena, MT: Falcon, 184 pp., 1995. (15604-43529) Shows the firestorm of political rhetoric and action over the 1995

wolf reintroduction program at Yellowstone National Park. Ranchers and residents warned of the dangerous wolves terrorizing their communities, while the animal rights people applauded the action loudly and vowed more species reintroduction in the future all over America. Reviewed at [egj.uidaho.edu/egj05/peek01.html]. How rarely does anyone find a middle ground! In the media and in arguments it is always the extremes that get touted publicly: all or nothing; kill all wolves or save all wolves. Rationality flies out the window in these kinds of turf wars.

707.Fitter, R. *Wildlife for Man: How and Why We Should Conserve Our Species.* London: Collins, 223 pp., 1987. How to monitor species populations, why the populations are dropping, and what we can do about it. See the U.S. Department of Fish and Wildlife page on wildlife conservation at [endangered.fws.gov/endspp.html].

708.Fontaubert, A. Charlotte de, Downes, David R., and Agardy, Tundi S. *Biodiversity in the Seas: Implementing the Convention on Biological Diversity in Marine and Coastal Habitats.* Gland, Switzerland: IUCN, 92 pp., 1996. (2831703387) In some ways, marine conservation is a trickier business than conservation of the land creatures. Who owns the oceans? No country, to be sure. Aside from small "national water" sections which extend only ten or twenty miles from a nation's coastline, the oceans are international. A similar IUCN work is titled *Migratory Species in International Instruments: an Overview* (1986). Who is responsible for protecting birds that fly through many countries? Read a large article about Australia's efforts to protect marine species at [www.environment.gov.au/life/general_info/biodivser_2/biod_5.html#HDR37].

709.Fossey, Dian. *Gorillas in the Mist.* Boston: Houghton Mifflin, 326 pp., 1988. (0395489288) This remarkable woman was the first to study mountain gorillas in their native habitat of Central Africa. Her project became more than a mere study when she attempted to protect the gorillas from poachers and zoo collectors. She was murdered, and the gorillas are now nearly extinct. This story was made into a fine movie starring Sigourney Weaver. See also Farley Mowat's biography of Fossey, *Woman in the Mists*, in this chapter. Find the website of the Gorilla Foundation at [www.gorilla.org/].

710.Freeman, Milton R., and Kreuter, Urs P. *Elephants and Whales: Resources for Whom?* Newark, NJ: Gordon & Breach, 321 pp., 1995. (2884490108) Issues on the conservation of large mammals (land and sea). Read an article by the author at [www.highnorth.no/po-be-an.htm].

711. Freese, Curtis H. *Harvesting Wild Species: Implications for Biodiversity Conservation.* Baltimore, MD: Johns Hopkins University, 704 pp., 1997. (080185573x) Fifteen international case studies linking sustainable development with biodiversity conservation. The author works for Montana State University, and is a researcher for the World Wildlife Fund. See a brief publisher review at [128.220.50.88/press/books/titles/s97/s97frha.htm].

712. Freese, Curtis H. *Wild Species as Commodities: Managing Markets and Ecosystems for Sustainability.* Washington, DC: Island, 256 pp., 1998. (1559635711) Argues that resources can only be protected if there are direct economic benefits to the local peoples. Based on a four-year study by the World Wildlife Fund. See a publisher review at [www.islandpress.com/islandpress/press/PR/wildspopr.html].

713. Gaston, K. J., and Spicer, J. I. *Biodiversity: An Introduction.* Cambridge: Blackwell Science, 128 pp., 1998. (0632049537) A very basic primer to explain the general questions of biodiversity concepts. See Gaston's faculty department (University of Sheffield) at [www.shef.ac.uk/~bmg/].

714. Gibson, Clark C. *Politicians and Poachers: The Political Economy of Wildlife Policy in Africa.* Cambridge: Cambridge University, 245 pp., 1999. The author studies three countries in Africa (Zambia, Kenya, and Zimbabwe) to show the "ins and outs" of wildlife protection (or non-protection) in the third world. Read one case study about the poaching of crocodiles in Africa at [www.american.edu/projects/mandala/TED/nilecroc.htm].

715. Gilbert, Oliver L., and Anderson, Penny. *Habitat Creation and Repair.* New York: Oxford University, 304 pp., 1998. (0198549679) The authors offer practical guidance on the little-discussed problem of fixing broken or missing habitats. Most of the books in this chapter (Animal Populations) simply state the myriad problems and disappearing homes of the animals; here is a book that tells how we might fix some of these problems. Of course, the creation and repair of habitats is not the preferable route: ideally we would have animals living in their own earthly realms as nature had provided. However, we are not in an ideal world, and we will have to learn how to create and fix new homes for the displaced species. Read a related article at [www.ijc.org/boards/wqb/hab_summ.html].

716. Gore, Al. *Earth in the Balance: Ecology and the Human Spirit.* Baltimore: Penguin, 407 pp., 1993. (0452269350) A strategic view of the rapid destruction and consumption of natural resources. The author was a senator from Tennessee and is now the vice president of the United States

under President Bill Clinton. Briefly reviewed at [www.fplc.edu/riskrevs/rv4.htm#ear]. Of course, opponents try to "debunk" much of the science quoted in this book.

717. Gray, Gary G. *Wildlife and People: The Human Dimensions of Wildlife Ecology*, reprint ed. Environment and the Human Condition series. Chicago: University of Illinois, 260 pp., 1995 (originally published in 1993). (0252063163) See a bibliography of resources on human attitudes toward wildlife at [www.dnr.cornell.edu/hdru/pubs/wildsttp.htm].

718. Green, Alan. *Animal Underworld: Inside America's Black Market for Rare and Exotic Species*. Washington, DC: Institute for Public Affairs, 336 pp., 1999. (1891620282) The author is an "investigative journalist" who says that even the captive-bred wild animals in the United States wind up being turned into consumer commodities like pelts or sport-hunting targets. This is the most eye-opening book I have seen on the problems with zoos and wildlife parks which commonly lie, cheat, and steal to avoid complying with federal and state regulations, and to hide embarrassing information from the general public. If you love zoos and do not wish to have your beliefs challenged, do not read this book! View the publisher promotion at [www.publicaffairsbooks.com/books/ani-sum.html] and find more information about each chapter at [www.animalunderworld.com/]. Reviewed at [www.seattletimes.com/news/entertainment/html98/anim_20000227.html].

719. Greenberg, Russell, and Reaser, Jamie. *Bring Back the Birds: What You Can Do to Save Threatened Species*. Mechanicsburg, PA: Stackpole, 312 pp., 1995. (0811725197) An ornithologist and a biologist explain how people can help local birds, help researchers, and take political action to protect birds. Long appendices on 159 bird species profiles, with their specific habitat needs. Find a bibliography on bird conservation at [www.si.edu/resource/faq/nmnh/birds.htm#CONSERVATION].

720. Groombridge, Brian, ed. *IUCN Red List of Threatened Animals*. Gland, Switzerland: International Union for Conservation of Nature and Natural Resources, 286 pp., 1996. (2831701945) Use the 1996 version online at [www.wcmc.org.uk/species/animals/animal_redlist.html].

721. Hadidian, John, et al., eds. *Wild Neighbors: The Humane Approach to Living with Wildlife*. Golden, CO: Fulcrum, 288 pp., 1997. (1555913091) Read an excerpt at [fulcrum-resources.com/html/wildexc.html]. See related questions and answers at [www.fund.org/facts/nuisance.html].

722. Hage, Wayne. *Storm Over Rangelands: Private Rights in Federal Lands*. Bellevue, WA: Free Enterprise, 265 pp., 1989. (0939571064) The author

says that there are some minimal property rights for individuals in the U.S. federal lands, but the federal government continually broadens its regulatory powers over that land. Read an article by the author at [www.libertymatters.org/GOLD%20Calf.html], and see [www.liberty matters.org/HISTORY.html]. Residents of the eastern United States have no idea how difficult land ownership can be west of the Mississippi River. The federal and state governments (together) own and control somewhere around 80 percent of the total land area. This means that residents of the western states must wrestle with all sorts of federal and state regulations, which are unknown to easterners. This is becoming a volatile issue in the west.

723. Halladay, Patricia, and Gilmour, D. A., eds. *Conserving Biodiversity Outside Protected Areas: The Role of Traditional Agro-Ecosystems.* Gland, Switzerland: IUCN, 227 pp., 1995. (2831702933) One major question for habitat protection is that of land ownership. By what right and what means do governments force landowners to restrict their usage of the lands in order to protect wildlife? This book (based on a 1994 workshop) looks at some of the areas where government regulations play almost no role in wildlife protection, and yet the habitats seem to be holding up well for the animal species living there: areas like mountainous South America, and Vietnam. Read a related online book, *Biodiversity Conservation in Transboundary Protected Areas*, at [books.nap.edu/bo_book/0309055768/html/index.html].

724. Hargrove, Eugene C., ed. *The Animal Rights, Environmental Ethics Debate*. SUNY Series in Philosophy and Biology. Albany: State University of New York, 273 pp., 1992. (0791409333) The author, who is the editor of *Environmental Ethics* magazine, tries to reconcile "deep ecology" with animal rights. Often, the protection of a whole ecosystem involves the destruction of many individual creatures and thus creates tension between the movements. See the author's resume at [www.cep.unt.edu/vech.html].

725. Harland, David. *Killing Game: International Law and the African Elephant*. Westport, CT: Praeger, 219 pp., 1994. (0275947998) The effects of the 1989 and 1992 bans on ivory. "First, can international law be an effective tool for the conservation of wildlife? And second, is international law serving the African elephant well?" (p. ix) The author is an expert at the United Nations on international law. Read the table of contents at [info.greenwood.com/books/0275947/0275947998.html]. The best ivory, for carving purposes, comes from elephants (not narwhals or walruses). For many centuries, elephants have been killed for their tusks to use in the decorative arts. The advent of rifles and vehicles made the hunting much simpler. Elephant populations have been decimated in the twentieth

century. Since the best ivory comes on the larger adult elephants, populations have become younger and less experienced. Also, the general poverty of African people has made poaching for ivory a very profitable business; the price of one set of tusks could finance a family for a year or more.

726.Heintzelman, Donald. *Wildlife Protectors Handbook: How You Can Help Stop the Destruction of Wild Animals.* Santa Barbara, CA: Capra Press, 157 pp., 1992. (0884963462) A practical guide for activists on how to make an immediate difference in saving habitats and species.

727.Hemley, Ginette, and Fuller, Kathryn S., eds. *International Wildlife Trade: A CITES Sourcebook.* Washington, DC: World Wildlife Fund, 180 pp., 1994. (1559633484) See an interview with Hemley at [www.emagazine.com/may-june_1999/0599conversations.html]. Read articles for and against CITES at [www.saep.org/subject/natcon/natcities.html] and [www.naiaonline.org/ 97cites1.html].

728.Hoage, R. J., ed. *Animal Extinctions: What Everyone Should Know.* Washington, DC: Smithsonian, 192 pp., 1985. (0874745217) Twelve essays based on speeches at the National Zoological Symposium for the Public in 1982. Specific problems and solutions are addressed. One essay is written by Paul Ehrlich. Find a related bibliography at [osu.orst.edu/Dept/ag_resrc_econ/biodiv/biblio_1_5.htm].

729.Hoyt, John A. *Animals in Peril: How "Sustainable Use" is Wiping Out the World's Wildlife.* Garden City: Avery, 257 pp., 1994. (0895296489) "'[S]ustainable use' — [is] an animal management concept that allows for legalized killing in the name of conservation . . . this theory is not working." (back cover) Alternatives to sustainable use. The author says that economics, not truth, is shaping the sustainable use theory. Read an excerpt at [lynx.uio.no/jon/lynx/peril.html] and an article by the author at [arrs.envirolink.org/ar-voices/intrinsic_value.html].

730.Humane Society of the United States. *Animal Sheltering Magazine* (periodical). Humane Society of the United States (HSUS), 1999. See the HSUS website at [www.hsus.org/programs/companion/animal_sheltering/index.html]

731.Iudicello, Suzanne, et al. *Fish, Markets, and Fishermen: The Economics of Overfishing.* Washington, DC: Island, 157 pp., 1999. (1559636424) Find a brief summary at [www.islandpress.org/books/bookdata/fishmarkfishmen.html]. The authors list many of the incentives that modern fishermen have to over-fish. Read several brief related articles at [forum.ra.utk.edu/for11-

2.htm]. It is difficult to comprehend why fishermen would overfish. If worldwide fish stocks are depleted, jobs will be lost by those fishermen. Perhaps the idea is that fish will be much more valuable when stocks are low. Maybe there is no thought at all, just the desire for immediate profits without consideration of the long-term consequences.

732. Jacobs, Lynn. *Waste of the West: Public Lands Ranching*. Tucson, AZ: Lynn Jacobs, 650 pp., 1992. (0962938602) The ecological damage of grazing cattle on land. Read the whole book online at [www.apnm.org/waste_of_west/index.htm]. Find an article by the author at [arrs.envirolink.org/ar-voices/eyeeye.html]. See an interesting site called Oust the Cows at [www.sw-center.org/swcbd/grazing/grazing.html].

733. Jakobsson, Kristin M., and Dragun, Andrew K. *Contingent Valuation and Endangered Species: Methodological Issues and Applications.* New Horizons in Environmental Economics series. Northampton, MA: Edward Elgar, 296 pp., 1996. (1858984645) The authors offer very concrete and definable ways to place monetary value on wildlife and ecological resources. This becomes important to politicians and to lawyers who must try to determine the value of something in order to protect it (or sue over it). See the table of contents at [www.hgur.se/envir/miljo21/littips/0092.html]. Use a very large bibliography on the subject of contingent valuations of resources at [www.sscnet.ucla.edu/ssc/labs/cameron/nrs98/cvinv.htm]. Read a substantial study about the economic valuation of wildlife at [www.puce.edu.ec/Investigacion/fatima/Whitep.htm].

734. Johnson, William. *The Monk Seal Conspiracy.* Iridescent Publishing, 1997? (0946097232) This species of seal lives only in a small area between Turkey and Greece, and is one of the 12 most endangered species in the world. Read the whole book online at [www.iridescent-publishing.com/MSC/msc_hub.html].

735. Kaiser, M. J., and Groot, S. J. de. *Effects of Fishing on Non-Target Species and Habitats.* Cambridge: Blackwell Science, 288 pp., 1999. (0632053550) Usually, conservation in regard to an animal-harvesting industry (like fishing) is discussed solely in relation to the "target species," that is, the type of fish that the fisherman are trying to catch. What is sometimes forgotten is that the removal of X type of fish has effects upon the other parts of the ecosystem. What about the fish that eat X fish? What about the organisms that X fish used-to-eat; will they suddenly have a population explosion because their predator has been eliminated? This book offers a broader view of the implications of "target species" decimation in relation to other species. There is another book on the same subject by Blackwell Science, called *The Effects of Fishing on Marine Ecosystems and*

Communities, by S. Hall. Read a lengthy discussion of overfishing at [www.island.net/~psa/greenpeace.htm].

736. Kellert, Stephen R. *Kinship to Mastery: Biophilia in Human Evolution & Development.* Washington, DC: Island Press, 256 pp., 1997. (1559633727) Humankind is more dependent upon its attachments to the animal world than it thinks, both spiritually and economically. Reviewed at [www.yale.edu/opa/newsr/97-10-13-01.science.html] and [asa.calvin.edu/ASA/book_reviews/12-98.htm#Kellert].

737. Kellert, Stephen R. *The Value of Life: Biological Diversity and Human Society.* Washington, DC: Island, 280 pp., 1997. (1559633174) Biodiversity is important not just because of economic or ecological consequences, but also because of social, emotional, and psychological impacts upon mankind. Includes a chapter on endangered species. Reviewed at [asa.calvin.edu/ASA/book_reviews/6-97.htm].

738. Knight, Richard L., and Gutzwiller, Kevin J., eds. *Wildlife and Recreationists: Coexistence through Management and Research.* Washington, DC: Island, 384 pp., 1995. (1559632577) Seven case studies of direct and indirect effects of recreationists on diverse wildlife in ecosystems. Bibliography at [www.tamu.edu/ethology/RefDistu.html]. See an article on the effects of snowmobiles on wildlife at [www.fund.org/facts/snowcomm.html]. Read an article in favor of recreation in parks at [www.outfittermag.com/articles/881post.html]. The snowmobile and boating arguments are where the vast differences between the preservationists (animal rights people) and conservationists (most others) show up the most clearly. The Clinton administration is pushing out the use of boats and snowmobiles in national parks, saying that the noise upsets the wildlife. Opponents say that the animals can live with some strange noises, and national parks are meant to be enjoyed. Preservationists are pushing for bans on all motor-vehicles in many parks, so that visitors must park outside and ride in on buses. Others are demanding bans on flights over the Grand Canyon and other parks.

739. Knoder, C. Eugene. *Overgrazing and Excessive Predator Control: Twin Abuses of the Public Lands.* Washington, DC: Audubon Society, 1989? Find grazing information at [www.orst.edu/instruction/bi301/wpubland.htm] and thoughts on coyote control at [texnat.tamu.edu/symp/coyote/p7.htm]. I could not find much information on this book, but the topic is important. When public lands are leased to ranchers, the government is responsible for predator control when ranchers say that predators are killing their cattle. The expense of predator control is immense; it would be

much cheaper to pay for the dead cattle than to hunt down the few remaining predators by helicopter.

740.Kohm, Kathryn A. *Balancing on the Brink of Extinction: The Endangered Species Act and Lessons for the Future*. Washington, DC: Island, 318 pp., 1991. (1559630078) See how animals get onto the Endangered Species Act list at [www.cnie.org/nle/biodiv-10.html].

741.Landau, Diane, and Stump, Shelley, eds. *Living with Wildlife: How to Enjoy, Cope with, and Protect North America's Wild Creatures around Your Home and Theirs*. San Francisco: Sierra Club, 341 pp., 1994. (0871565471) Wildlife can be a nuisance to people at times. They sometimes nibble on gardens and flowers, dig in garbage cans; or nest beneath homes. There are nondeadly ways of reducing animal damage to property.

742.Lawton, John H., and May, Robert M., eds. *Extinction Rates*. New York: Oxford University, 248 pp., 1995. (019854829x) This is a rather grim but important subject. Correct determination of species extinction may lead to more proper protections and solutions, whereas incorrect determinations (exaggerations) may make headlines, but ultimately prove counter-productive if policy makers come to distrust the "Chicken Little" variety of statistical analysts. See an article about extinctions at [oneworld.org/ni/issue288/keynote.html].

743.Laycock, George. *The Alien Animals: The Story of Imported Wildlife*. New York: Ballantine, 240 pp., 1970. The dangers of importing nonnative species. There have been a few successes, like rabbits and pheasants, but far more failures, where the new species destroys the ecosystem (no predators to kill them). This book was later condensed into *Animal Movers*. The author has written several books and articles on wildlife. See an article about exotic species introductions at [darwin.bio.uci.edu/~sustain/bio65/lec09/b65lec09.htm].

744.Libecap, Gary D. *Locking up the Range: Federal Land Controls and Grazing*, reprint ed. San Francisco: Pacific Research Institute for the Public, 109 pp., 1985 (originally published in 1981). (0884103870) See a related article at [www.zianet.com/wblase/endtimes/rebel2a.htm]. Read the author's resume at [nber.nber.org/vitae/vita340.htm]. Find a bibliography on property rights at [encyclo.findlaw.com/biblio/1910.htm].

745.Liddle, M. *Recreation Ecology: The Ecological Impact of Outdoor Recreation*. Boston: Kluwer Law International, 664 pp., 1998. (04122-6630x) Find a bibliography and links at [www.unbc.ca/rrtdo/reference/

ref18.html], and read an article at [www.fw.vt.edu/forestry/cpsu/rececol.html].

746.Littell, Richard. *Endangered and Other Protected Species: Federal Law and Regulation*. Washington, DC: BNA Books, 613 pp., 1992. (08717-9747x) Nineteen federal laws and court cases.

747.Long, John ed. *Attacked! By Beasts of Prey and Other Deadly Creatures: True Stories of Survivors*. New York: McGraw-Hill, 192 pp., 1997. (0070386994) Seventeen stories of attacks by various creatures including lions, alligators, sharks, tigers, bears, and elephants. See a bibliography of animal attacks at [www.igorilla.com/gorilla/animal/books.html]. Animal attacks are quite rare in the United States because most predators have been wiped out, and are found only in zoos. Attacks are more common in other continents like Africa and Asia; but still are relatively rare. Animals generally do not attack humans; we are fairly large, make a lot of noise, and do not taste especially good. Only desperate beasts actually learn to eat people, but such stories are exciting and make for great newspaper stories and horror movies.

748.Lowe, D., et al., eds. *The Official World Wildlife Fund Guide to Endangered Species of North America*, revised ed. Austin, TX: Beacham, 1258 pp., 1998 (originally published in 1990). (0933833466) This two-volume set lists 547 endangered North American species. See a bibliography on the subject at [www-mmc.library.wisc.edu/college/roadmaps/endspecies.htm].

749.Lutts, Ralph H., et al. *The Wild Animal Story*. Animals, Culture, and Society series. Philadelphia: Temple University, 328 pp., 1998. (15663-95933) An "exploration of the popular genre of wild animal stories," and what these tales show about society's views on wildlife. Find a brief author biography at [www.goddard.edu/offcamp/bama/facbios.htm].

750.Lyster, S. *International Wildlife Law*. Cambridge: Grotius, 470 pp., 1985. (0906496462) Read a bibliography at [www.personal.psu.edu/staff/g/a/gap6/IntlEnv/biblio.html]. One difficulty with international laws of any sort is enforcement. Who will enforce sanctions or penalties on a nation for violations of wildlife laws? Public shame through media outlets is about the only effective tool in our present system. National-level sanctions might be useful, but sanctions are not even forthcoming on major human rights violations (like China), so the expectation or hope that animal rights issues would make waves is unrealistic in this system. Many folks do not believe that a world court can be effective or should even exist, overruling national sovereignty.

751.MacE, Georgina M., et al., eds. *Biotechnology and the Conservation of Genetic Diversity: The Proceedings of a Symposium Held at the Zoological Society of London*. Oxford: Clarendon, 240 pp., 1992. (0198540302) MacE works with the Institute of Zoology in London.

752.Makombe, Kudzai, ed. *Sharing the Land: Wildlife, People and Development in Africa*. Washington, DC: International Union for the Conservation of Nature and Natural Resources, 1994. (2831701937) Read this book online at [art.org.uk/articles/art_sharing.html].

753.Mann, Charles C., and Plummer, Mark L. *Noah's Choice: The Future of Endangered Species*. New York: Knopf, 302 pp., 1995. (0679420029) Read a negative review at [www.arktrust.org/animals/books/noahchoi.html] and positive reviews at [www.discovery.org/w3/discovery.org/noahschoice.html] and [egj.lib.uidaho.edu/egj06/tobin.html].

754.Manning, Laura L. *The Dispute Processing Model of Public Policy Evolution: The Case of Endangered Species Policy Changes from 1973 to 1983*. New York: Garland, 490 pp., 1990. (0824025202) This topic is of interest because it emphasizes the odd nature of our current Endangered Species Act system of protection. There is a very convoluted political system to be harnessed in putting a species on or taking a species off of the protected list. Lawyers are necessary to use this system with any hope of success. See a related article at [www.uidaho.edu/cfwr/pag/pag13es.html].

755.Matthews, Anne. *Where the Buffalo Roam*. New York: Grove Weidenfeld, 189 pp., 1992. (0802114083) A controversial plan to return 25 percent of the land area of the U.S. Great Plains to original prairie, so as to restore the buffalo and other indigenous species. Of course, many midwesterners oppose the idea. See her faculty page at [www.nyu.edu:81/gsas/dept/journal/Faculty/bios/matthews.htm]. This book was one of the finalist nominees for a Pulitzer Prize.

756.McCullough, Dale R., ed. *Metapopulations and Wildlife Conservation*. Washington, DC: Island, 432 pp., 1996. (1559634588) This is an interesting concept, if perhaps untested and theoretical. If I understand it correctly, the idea is that lots of distant and separate "isolated patches" of species' populations can be sustained more easily than having huge, untouched wilderness habitats. If true, then the U.S. system of national and state parks might be a feasible and workable system of protection for endangered species. See the author's page at [www.cnr.berkeley.edu/~mcculla/].

757.McShea, William J., et al., eds. *The Science of Overabundance: Deer Ecology and Population Management*. Washington, DC: Smithsonian, 432 pp., 1997. (1560986816) See articles on the subject at [www.naiaonline.org/pred1.html], [lutra.tamu.edu/dms/backgnd.htm], and [www.arec.umd.edu/policy/Deer-Management-in-Maryland/warren.htm]. Hunters claim that hunting is good because it culls off the surplus population, which might otherwise starve. Opponents claim that deer populations in particular are kept artificially high so that the excess can be hunted.

758.Meacham, Cory J. *How the Tiger Lost Its Stripes: An Exploration into the Endangerment of a Species*. New York: Harcourt Brace, 288 pp., 1997. (0151002797) See a bibliography at [www.escati.com/tigers_conservation_books.htm], and articles at [www.perc.org/ps12.htm] and [www.earthtimes.org/oct/environmentthewildtigerfacesoct18_98.htm].

759.Mighetto, Lisa. *Wild Animals and American Environmental Ethics*. Tucson: University of Arizona, 209 pp., 1991. (0816511608) How Americans have perceived wild creatures. What arguments were used to protect animals? History of the changing attitudes over the last century, showing an interesting progression of ideas. Societal, cultural, and sociological reasons for the various harms and helps to wildlife conservation.

760.Morris, Desmond. *The Animal Contract: Sharing the Planet*. Virgin, 169 pp., 1990. (1852273925) Based on the TV series. The author says that man is too successful for the good of the planet and we will soon see the consequences. Read a brief biography of the author at [www.sirc.org/about/desmond_morris.html]. Find a bibliography of conservation at [ci.moscow.com/naturesown/book_cons.html].

761.Morrison, Michael L., Marcot, Bruce G., and Mannan, R. William. *Wildlife-Habitat Relationships: Concepts and Applications*, 2nd ed. Madison: University of Wisconsin, 435 pp., 1998. (0299156400) A scholarly work for advanced studies in measurement of wildlife populations and habitats. Basis for understanding ecology and sound ecological research methods. See the publisher's page with links to a few updated sections at [gopher.adp.wisc.edu/wisconsinpress/books/0182. html].

762.Moulton, Michael, and Sanderson, James. *Wildlife Issues in a Changing World*, 2nd revised ed. Boca Raton, FL: St. Lucie, 368 pp., 1997. (15744-40683) Read the table of contents and some biographical information at [www.crcpress.com/index.htm?catalog/L1351]. This book is commonly used as a textbook in ecology classes.

763. Mowat, Farley. *Never Cry Wolf*, reprint ed. Mattituck, NY: Amereon Ltd., 164 pp., 1988 (originally published in 1957). (0891908234) Funny, insightful, and wonderful true story about a young scientist sent to study wolves in the Arctic. He came to the unpopular conclusion that the Caribou herds were not being ravaged by wolves, as theorized, but by human trophy hunters. Mowat may be the most popular Canadian nonfiction author. See Mowat's bibliography at [myunicorn.com/bibl6/bibl0658.html]; and read an article about arctic wolves at [www.geocities.com/Heartland/Woods/3507/arctic wolf.html].

764. Mowat, Farley. *Woman in the Mists: The Story of Dian Fossey and the Mountain Gorillas of Africa*. New York: Warner, 380 pp., 1987. (04465-13601) Dian Fossey was with gorillas for 19 years (see her own book, *Gorillas in the Mist*). Mowat wrote this comprehensive biography with the help of Fossey's journals, letters, and interviews with friends and coworkers. See a list of books about Fossey at [www.waypt.com/users/~gorillas/dian.htm], and an article about endangered gorillas at [www.kilimanjaro.com/gorilla/marcel.htm]. This book shows some of the dark side of conservation efforts, as Fossey took up direct-action methods against the local peoples out of desperation to save the gorilla groups. In order to protect a species from poachers, how far will we go? Do we kill the poachers? Kidnap their children? Publicly humiliate the government by writing articles on their incompetence and corruption? I am not saying that these are necessarily wrong, but they are not the comfortable animal rights conversations we have in our cozy homes. When it comes right down to it, someone may have to use physical force to protect a creature from death against determined enemies. Who decides when and how to pull the trigger? Us? Dian Fossey? The government of Zaire or Congo?

765. Musgrave, Ruth S., et al. *Federal Wildlife Laws and Related Laws Handbook*. Rockville, MD: Government Institutes, 665 pp., 1998. (086587557x) See statute summaries at [ipl.unm.edu/cwl/fedbook/statute_frame.htm].

766. Musgrave, Ruth S., and Stein, Mary Anne. *State Wildlife Laws Handbook*. Rockville, MD: Government Institutes, 840 pp., 1993. (0865873577) Find brief state law information at [www.defenders.org/bio-st00.html].

767. Myers, Norman. *The Sinking Ark: A New Look at the Problem of Disappearing Species*. Oxford: Pergamon, 307 pp., 1987 (originally published in 1979). (0080245013) The author has been criticized for overly dire predictions: he said we would lose one million species in the last 20 years of this century. He emphasizes the protection of habitats and not just specific species. He says that the rate of extinction is attributable to man's

exploitation of resources. Read an article by the author at [www.oneworld.org/patp/pap_7_4/myers.htm].

768.National Academy of Sciences. *Decline of the Sea Turtles: Causes and Prevention*. Washington, DC: National Academy Press, 280 pp., 1990. (030904247x) Threats to the populations and overview of the species, including habits, migration, and conservation methods. Read the whole book online at [books.nap.edu/catalog/1351.html]. The main cause of the sudden demise of sea turtle populations is habitat loss and human hunting. The turtles nest on the beaches of Latin America to lay their eggs, but poachers wait on the beach and catch the adults coming up the beach, to eat their meat and sell the shells. There is practically no enforcement of wildlife laws in Latin America. Furthermore, hotels and resorts, a great moneymaker for poor countries, like to locate on the beaches, and the nesting sites are being turned into "tourist traps."

769.National Academy of Sciences, Committee on Porpoise Mortality from Tuna Fishing. *Dolphins and the Tuna Industry*. Washington, DC: National Academy Press, 176 pp., 1992. (0309047358) Analysis of the scientific questions regarding dolphin deaths by tuna fishermen. Overview of U.S. laws and policies. Offers possible solutions: recommending education, and new fishing gear designed to reduce mortality. See the table of contents at [books.nap.edu/catalog/1983.html]. Read the Tuna Conventions Act at [ipl.unm.edu/cwl/fedbook/tunaconv.html]. See a related article at [www.cnie.org/nle/mar-14.html]. Apparently it is common for net-fishermen to target pods of dolphins because dolphins are known to swim over schools of tuna. An outcry in the 1980s led to a great reduction in the allowable number of "incidental kills" of dolphins in the nets. There is some debate as to whether the rules are being enforced or followed.

770.National Research Council. *Animals as Sentinels of Environmental Health Hazards*. Washington, DC: National Academy Press, 176 pp., 1991. (0309040469) Observing domestic and wild animals to determine dangerous conditions (like a caged canary in a mineshaft), using natural populations, or by taking animals to the area intentionally for testing. Some scientists view the decimation and mutation of frog populations as a sign of impure water or atmosphere. See the table of contents at [books.nap.edu/catalog/1351.html].

771.Nilsson, Greta. *The Endangered Species Handbook*. Washington, DC: Animal Welfare Institute, 261 pp., 1990 (originally published in 1983). (0938414097) Includes class projects. See an updated list of animals on the Endangered Species Act at [www.cnie.org/nle/biodv-18.html].

772. Norton, Bryan G., ed. *Preservation of Species: The Value of Biological Diversity*. Princeton, NJ: Princeton University, 318 pp., 1988 (originally published in 1986). (0691024154) Read a similar book online at [pompeii.nap.edu/catalog/catalog.cfm?record_id=1925]. See an article proposing a constitutional amendment on biodiversity at [198.240.72.81/bio-co00.html].

773. Noss, Reed F., and Cooperrider, Allen Y. *Saving Nature's Legacy: Protecting and Restoring Biodiversity*. Washington, DC: Island, 443 pp., 1994. (155963247x) The authors are conservation biologists who work for Defenders of Wildlife. Read a seven-part article by Noss at [www.defenders.org/amee01.html]. See also a good article and bibliography at [www.gn.apc.org/eco/resguide/2_7.html].

774. Noss, Reed F., O'Connell, Michael A., and Murphy, Dennis D. *The Science of Conservation Planning: Habitat Conservation Under the Endangered Species Act*. Washington, DC: Island, 272 pp., 1997. Three leading biologists discuss the planning of successful conservation efforts. See a related article at [thecity.sfsu.edu/users/IEH/html/conservation_planning.html].

775. Olney, P. J. S., et al. *Creative Conservation: Interactive Management of Wild and Captive Animals*. London: Chapman & Hall, 1994. (04124-95708) Based on a 1992 conference on the role of zoos in global conservation. More than 30 papers on populations, returning captive animals to wild habitats, and breeding endangered species in captivity.

776. Olsen, Jack. *Slaughter the Animals, Poison the Earth*. New York: Simon & Schuster, 287 pp., 1971. (0671209965) How ranchers and others spread poisons and traps to kill predators, like coyotes, to protect sheep and cattle. Many of these poisons and traps end up killing pets, harmless creatures, and humans. Read a publisher review at [www.mjq.net/jackolsen/slaughter.htm]. Read an article on the violence inherent in changing a habitat at [egj.lib.uidaho.edu/egj09/shelton1.html]. This is one reason for opposition to the leasing of public lands in the western United States for cattle ranching. To keep the cows safe, predators are killed. The grasses eaten by the cows cannot then be eaten by native species like bison, or elk, or deer.

777. Overrein, Lars N., et al. *Acid Precipitation: Effects on Forest and Fish*, 2nd ed. Oslo, Norway: Norwegian Council on Science and Industrial Research, 1981. Read an article on the subject at [www.enviroliteracy.org/acid_rain.html], and find a bibliography at [www.gn.apc.org/eco/resguide/1_4vi.html].

778.Pain, Deborah J., et al., eds. *Farming and Birds in Europe: The Common Agricultural Policy and its Implications for Bird Conservation*. New York: Academic Press, 436 pp., 1997. (0125442807) Read a related article at [www.mluri.sari.ac.uk/~mi361/tdv/pienkowski.htm].

779.Palmer, James L. *Game Wardens vs. Poachers: Tickets Still Available,* 5th ed. Iola, WI: Krause, 160 pp., 1993. (0873412184) A humorous look at the adventures of a game warden.

780.Parker, Ian S. C. *Oh Quagga! Thoughts on People, Pets, Loving Animals, Shooting Them, and Conservation*. I. Parker, 1983. The Quagga was a zebra-like creature that was hunted to extinction in the 1880s in southern Africa. Read an article about public opinion on species conservation at [www.umich.edu/~esupdate/library/97.05-06/czech.html].

781.Peacock, Donald. *People, Peregrines and Arctic Pipelines: The Critical Battle to Build Canada's Northern Gas Pipelines*. Seattle: University of Washington, 1980. (0888941382) See an article about the politics of endangered species conservation at [egj.lib.uidaho.edu/egj09/olaugh1. html]. This Canadian incident would be roughly equivalent to the current animal rights and environmental opposition to oil well exploration in Alaska or off the coasts of the United States. One side says the animals will all die, the other side says the animals will be fine.

782.Peterson, Dale. *The Deluge and the Ark: A Journey into Primate Worlds*. New York: Avon, 378 pp., 1989. (0380711990) The author, a professor at Tufts University, traveled to many countries to see endangered monkeys and their habitats. Read a brief biography of the author at [www. bkstore.com/tufts/fac/peterson.html].

783.Pierce, Christine, and Vandeveer, Donald, eds. *People, Penguins, and Plastic Trees: Basic Issues in Environmental Ethics*, 2nd ed. Belmont, CA: Wadsworth, 267 pp., 1995 (originally published in 1986). (0534179223) A collection of essays about animals, anthropomorphism, extinction, habitats, and ecosystems.

784.Prescott-Allen, Christine and Robert ed. *Assessing the Sustainability of Uses of Wild Species: Case Studies and Initial Assessment Procedure*. Gland, Switzerland: IUCN, 137 pp., 1996. (2831702879) How to assess the possibilities and formulate policies for the “harvest” of wild animals. The IUCN publishes a number of species-specific guides to conservation, including books on elephants, dolphins, whales, wild horses, and many other types of creatures. See their website at [www.iucn.org] for a complete listing. IUCN books are distributed in the United States by Island Press [www.islandpress.org].

785. Prescott-Allen, Christine, and Allen, Robert. *The First Resource: Wild Species in the North American Economy*. New Haven, CT: Yale University, 529 pp., 1986. (0300032285)

786. Pringle, Laurence. *Feral: Tame Animals Gone Wild*. New York: Macmillan, 110 pp., 1983. (0027754200) Written for juveniles. See a related site at [www.feralcats.com/faq.htm]. Feral animals are those creatures that were raised as pets or domestic animals, but for some reason returned to the wild lifestyle. Usually this is not by choice, but because an owner decided to move to the city and drops the pet off in the forest rather than finding another home for it. Feral dogs sometimes become pack animals like their wolf ancestors and are more dangerous, since they are not afraid to approach homes and kill pets or domestic animals for food.

787. Reaka-Kudla, Marjorie L., Wilson, Don E., and Wilson, Edward O., eds. *Biodiversity II: Understanding and Protecting our Biological Resources*. Washington, DC: Joseph Henry Press, 560 pp., 1997. (0309052270) More than thirty essays on "strategies necessary to stem the tide of this mass extinction event." (back cover) "[W]e may soon be bereft of species that could help us fight disease and produce useful products — not to mention bringing us the wonder of natural life." (dust jacket flap) Read the whole book online at [pompeii.nap.edu/ catalog/catalog.cfm?record_id=4901]. See a bibliography on biodiversity at [ecoethics.net/bib/tl-066-a.htm].

788. Rogers, Raymond A. *The Oceans are Emptying: Fish Wars and Sustainability*. Montreal, Canada: Black Rose Books, 176 pp., 1995. (155-1640317) "Although this inquiry considers a wide range of issues about the relationship between human society and nature, I always return to the question of why the hooks come up empty. It is a fisherman's question." (p. 1) The author was a full time Nova Scotian fisherman for more than a decade, and now is a university teacher.

789. Rohlf, Daniel J. *The Endangered Species Act: A Guide to Its Protections and Implementation*. Stanford, CA: Stanford Environmental Law Society, 207 pp., 1989. (0942007336) See an article on improving the Endangered Species Act at [www.edf.org/pubs/Reports/help-esa/]; and an article on the future of such laws at [cali.kentlaw.edu/student_orgs/lawrev/text69_4/ Kieter.htm].

790. Rowan, Andrew N., ed. *Animals and People Sharing the World*. Hanover, NH: University Press of New England, 192 pp., 1988. (0874514495) Find

an article and bibliography about populations at [www.gn.apc.org/eco/resguide/1_10ii.html].

791. Ryder, Richard Dudley, ed. *Animal Welfare and the Environment: An RSPCA Book.* London: Duckworth, 216 pp., 1992. (0715624032) Papers from an RSPCA conference in England. Is there a conflict between the interests of individual animals and the whole species or ecosystem? What about culling of animal populations? What about the rights of indigenous peoples to hunt? Reviewed at [www.psyeta.org/sa/sa3.2/ryder.html] and [www.newscientist.com/nsplus/insight/animal/ecology.html]. Read an extensive article on the subject at [www.carleton.edu/curricular/ENTS/faculty/dale/dale_animal.html].

792. Schaller, George B. *The Last Panda.* Chicago: University of Chicago, 291 pp., 1993. (0226736288) This author was allowed into China to make the first detailed study of wild pandas. Can the species survive, with less than 1000 left in the wild? The pandas are worth more than $100,000 alive, and $10,000 dead on the black market. A four-year study by the director of science at Wildlife Conservation International. Hear a RealAudio interview with the author online at [www.npr.org/ramarchives/nf6o1101-3.ram]. See an article about endangered pandas at [www2.deasy.psu.edu/rps/mar94.panda.html].

793. Seshadri, Balakrishna. *Call of the Wild: Survival in the Sun.* India's Wildlife and Wildlife Reserves series. New Delhi, India: Sterling Publishing Private Ltd., 1996. (8120716124)

794. Shafer, Craig L. *Nature Reserves: Island Theory and Conservation Practice.* Washington, DC: Smithsonian, 208 pp., 1991. (0874743842) Read an article on the subject at [fp.bio.utk.edu/bio250/jamie/island_biogeography.html] and a bibliography at [osu.orst.edu/Dept/ag_resrc_econ/biodiv/biblio_2_6_2.htm].

795. Sherry, Clifford. *Endangered Species: A Reference Handbook.* Contemporary World Issues series. Santa Barbara, CA: ABC-CLIO, 269 pp., 1998. Written for young adults. (0874368103) Read a large article about endangered species at [www.gsenet.org/library/23wld/SPECIES-.TXT].

796. Shogren, Jason F., ed. *Private Property and the Endangered Species Act: Saving Habitats, Protecting Homes.* Austin: University of Texas, 176 pp., 1999. (029277737) See related articles at [www.uwyo.edu/enr/ienr/esaprin.thm], [heritage.org/library/ backgrounder/bg1234.html], and an article with bibliography at [www.gn.apc.org/eco/resguide/2_20.html]. Many landowners have become angry at the U.S. and state governments over

draconian enforcement of Endangered Species Act interpretations. Many of the interpretations are based on very vague definitions of what defines a species. Once a creature is defined as endangered, no development is permitted on the habitat land, which means that the landowners are unable to build upon it. Some of the species that are called "endangered" are not the famous and cuddly creatures, but insects, fish, or even plants. Many landowners believe that their property is more important than some oddly colored wasp.

797. Sigler, William F. *Wildlife Law Enforcement*, 4th ed. New York: McGraw-Hill, 342 pp., 1995. (0679202690) See a bibliography on this topic at [www.fw.umn.edu/FW5603/5603ref.HTML].

798. Simberloff, Daniel, et al., eds. *Strangers in Paradise: Impact and Management of Nonindigenous Species in Florida.* Washington, DC: Island, 480 pp., 1997. (1559634308) The author is a biology professor at Florida State University. Read about one agency that is fighting the problem of bio-invasion, at [anstaskforce.gov/accomp.htm]. Florida is especially susceptible to invasion by non indigenous species because it is a popular tourist and shipping area. Many bio invaders are borne on foreign cargo ships. Even if these creatures do not stow away aboard ship in a crate; they may hitch a ride inside a ship's ballast tanks. Since the creatures are new to the area, there may be no predators to control their population. They may wipe out many native species in this new friendly environment.

799. Speight, M. R., et al. *Ecology of Insects.* Cambridge: Blackwell Science, 360 pp., 1999. (0865427453) Perhaps because insects are rather unpopular creatures, the animal rights people have not spend much energy lobbying to protect them from extinction. Since insects make up a majority of Earth's species, however, they are probably vanishing at a greater rate than other creatures.

800. Sterba, James P., ed. *Earth Ethics: Environmental Ethics, Animal Rights, and Practical Applications*, 2nd ed. Englewood Cliffs: Prentice Hall, 2000. (013014827x) Not yet printed. Essay authors include Dave Foreman, Paul Ehrlich, and Eugene Hargrove. See the table of contents for the second edition at [www.phptr.com/ptrbooks/hss_013014827x].

801. Stretch, Mary Jane. *For the Love of Wild Things: The Extraordinary Work of a Wildlife Center.* Harrisburg, PA: Stackpole, 150 pp., 1995. (08117-30158) See a sample wildlife center site at [www.pacificwildlife.org/who.html]. I visited a wildlife center on the Florida Gulf Coast recently. These facilities usually care for abandoned or wounded native species, often waterbirds. They often get birds that were entangled in fishing line, were

hit by automobiles, or were just found sick. After medical care, the animals are usually released, if they are thought to be capable of surviving again in the wild. Often these centers are staffed by volunteers and are nonprofit organizations.

802. Stuart, Chris and Tilde. *Africa's Vanishing Wildlife*. Washington, DC: Smithsonian, 208 pp., 1996. (1560986786) See an article about CITES and the wildlife trade at [www.nesl.edu/ annual/vol3/cite.htm]. Read an article about the Maasai people of Africa and wildlife at [www.montelis.com/ satya/backissues/dec97/maasai.html]. Africa has many major problems that contribute to the decimation of wildlife populations. But the truth of the matter is that African nations cannot view animal habitat problems as major priorities. Most of these nations are engulfed in bloody civil wars, plagued by famines and droughts, and lack basic medical care. The animal populations are rather low on their list of priorities. For people in the wealthy Western nations, however, it is fairly simple to say "you should protect those animals." There must be some real understanding of the problems faced by the native peoples, and larger fixes than simple public demands for bigger national parks in Africa.

803. Sugg, Ike, and Kreuter, Urs. *Elephants and Ivory: Lessons from the Trade Ban*. Institute of Economic Affairs, 1994. (0255363427) See several links and articles at [environment.miningco.com/library/weekly/bleph.htm]. Is the ivory trade ban a success? Can it be? In some ways it is rather like the American drug trade. Some people say we should legalize it, then prices will fall, and there would be less violence and less money to be made in it. Others say that is ludicrous, we should just fight harder. The same arguments are found in the ivory trade. Some say that by banning it, we have made ivory far more valuable, and thus the risks are worth taking to get it.

804. Takacs, David. *The Idea of Biodiversity: Philosophies of Paradise*. Baltimore, MD: Johns Hopkins University, 393 pp., 1996. (0801854008) Read a biography of the author at [www.monterey.edu/academic/institutes/ essp/staff/takacs.html]. Do animal rights people and environmentalists have a realistic view of the world, as to what it might be? Some critics say that we cannot return to Eden; we cannot save every creature. Even if this is true, shouldn't we try? Or is it simply evolution, "nature red in tooth and claw," fitting that some weaker species die off in favor of the stronger? Some of the more radical folks propose reintroducing Bubonic Plague and other nasty diseases to kill off 80 percent of the world's human population so that humans will again be "at harmony" with the Earth. Is that right, to murder millions of people so that more animals might survive? Does this

mean that the animals are better than the people? Is this paradise: a world with very few people?

805. Taylor, Victoria J., and Dunstone, N., ed. *The Exploitation of Mammal Populations.* New York: Chapman and Hall, 415 pp., 1996. (0412644207) Ecotourism, hunting, sustainable use issues. Edited papers from a conference by Universities Federation for Animal Welfare. Read related articles by Peter Singer at [www.animalliberation.org.au/comintro.html].

806. Thomas, Heather Smith. *The Wild Horse Controversy.* Cranbery: A.S. Barnes Co., 284 pp., 1979. (0498021912) Read the Wild Horses Act at [ipl.unm.edu/cwl/fedbook/wildhors.html]. See a related article at [www.fund.org/fact/ nvhorses.html].

807. Thorpe, J. E., et al., eds. *Conservation of Fish and Shellfish Resources: Managing Diversity.* New York: Academic, 206 pp., 1995. (0126906858) See a brief description of shellfish at [www.geocities.com/RainForest/Vines/5726/shell.htm]. Read an article about the harvesting of shellfish (more from an anthropologist's eyes than a biologist's, perhaps) at [www.library.arizona.edu/ej/jpe/volume_2/ascii-meltzoff.txt].

808. Tobias, Michael. *Nature's Keepers: Wildlife Poaching in the U.S. and Efforts to Stop It.* New York: Wiley, 304 pp., 1998. (0471157287) The author has written more than 25 books and made over 100 films. Read an interview with the author at [www.cpsweb.com/tobias.htm]; see links and a biography at [www.prairienet.org/community/clubs/youthtopia/tobiasbb.html].

809. Trexler, M., and Kosloff, L. *The Wildlife Trade and CITES: An Annotated Bibliography for the Convention on International Trade in Endangered Species of Wild Fauna and Flora.* Washington, DC: World Wildlife Fund, 346 pp., 1988. 2,300 entries. Read a related article at [www.hawaiilawyer.com/articles/jzbart2.htm]. See links at [www.animalwelfare.com/wildlife/wild-idx.htm].

810. Twiss, John R., Jr., and Reeves, Randall R., eds. *Conservation and Management of Marine Mammals.* Washington, DC: Smithsonian Institution Press, 496 pp., 1999. (1560987782) Thirty-one essays on the history and future of seals, whales, fisheries, and the ocean ecosystem. See a site on careers in working with marine mammals at [www.une.edu/cas/dls/marbio/marmam.html]. Read an entire online book, *Restoring and Protecting Marine Habitat: The Role of Engineering and Technology*, at [www.nap.edu/books/0309048435/html/index.html].

811. Varner, Gary E. *In Nature's Interests? Interests, Animal Rights, and Environmental Ethics.* New York: Oxford University, 166 pp., 1998. (0195108655) The author defends a "sentientist principle" that conscious beings have priority over unconscious beings, and, thus, humans have more interests than animals do. He concludes, however, that environmental goals can still be reached under this philosophy. Read articles by the author at [snaefell.tamu.edu/~gary/awvar/index.html]. See a biography and resume at [phil-www.tamu.edu/Faculty/Gary/vita.html].

812. Vaughan, Ray. *Endangered Species Act Handbook.* Rockville, MD: Government Institutes, 165 pp., 1994. (0865873925) See the author's biography at [www.wildlaw.org/ray.html]. Read the complete text of the Endangered Species Act at [www.sw-center.org/swcbd/activist/esatext.html]. Read an online book, *Science and the Endangered Species Act*, at [www.nap.edu/bo_book/0309052912/html/index.html].

813. Wagner, Frederic H., et al. *Wildlife Policies in the U.S. National Parks.* Washington, DC: Island, 251 pp., 1995. (1559634057) The author is a professor at Utah State University. Read one of his articles at [www.webcom.com/gallatin/Buffalo/Wagner.html].

814. Wagner, Frederick H. *Wild and Free-Roaming Horses and Burros.* Washington, DC: National Academy Press, 1982. See related articles at [www.fund.org/fact/wildhors.html] and [arrs.envirolink.org/news/blm_story.html].

815. Wallace, Richard L., compiler. *The Marine Mammal Commission Compendium of Selected Treaties, International Agreements, and Other Relevant Documents on Marine Resources, Wildlife, and the Environment.* Upland, CA: Diane, 3,547 pp., 1994. (0160493161) Read the Marine Mammal Protection Act at [ipl.unm.edu/cwl/fedbook/mmpa.html]. Related bibliography and links at [lib.law.washington.edu/ref/intsea.htm].

816. Ward, Peter. *The End of Evolution: A Journey in Search of Clues to the Third Mass Extinction Facing Planet Earth.* New York: Bantam, 302 pp., 1995. (0553374699) This work is a study of extinction from modern times to the present. The author says that the influence of man is bringing a new stage of extinctions. See an article at [members.aol.com/trajcom/private/endspcs.htm] and a bibliography at [www.cnw.com/~mstern/evolut_extinct.html].

817. Wenzel, George. *Animal Rights, Human Rights: Ecology, Economy and Ideology in the Canadian Arctic.* Toronto: University of Toronto, 206 pp., 1997 (originally published in 1991). (0802059619) Analyzes the impact of

animal rights campaigns against sealing on the Inuits and native peoples. Should subsistence hunting be banned as well as commercial hunting? Because they now use rifles instead of spears, and motorboats rather than canoes, are they still practicing native culture? "The seal controversy typifies the willingness of the animal rights movement to exercise a self-ascribed moral imperative toward Inuit and other aboriginal peoples." The author is a geography professor at McGill University. Reviewed at [www.ecotopia.be/ecotop/pubs/radicalecology/indiants.html/dark1.html].

818. White, Jan, and Frink, Lynne, eds. *Effects of Oil on Wildlife: Research, Rehabilitation and General Concerns.* Suisun City, CA: International Wildlife Rehabilitation, 210 pp., 1991. (1884196012) I have only found sketchy information on this book, so the above information is tentative. This work may be based on a conference about lessons learned from the Exxon Valdez spill. See a good article on a similar case at [www.zetnet.co.uk/sigs/braer/Braer.html], a bibliography on wildlife rehabilitation at [www.wildcare.com/books.htm], and a specific site on the spill restoration of the Exxon Valdez at [www.oilspill.state.ak.us/].

819. Wieland, Robert. *Why People Catch Too Many Fish.* Washington, DC: Center for Marine Conservation, 1992. See the publisher's site at [www.cmc-ocean.org/index.php3]. Read a related book at [pompeii.nap.edu/catalog/catalog.cfm?record_id=6335]. See a good site at [ehpnet1.niehs.nih.gov/docs/1996/104-4/focusocean.html].

820. Wigan, Michael. *The Last of the Hunter Gatherers: Fisheries Crisis at Sea.* Shrewsbury, England: Swan Hill Press, 270 pp., 1998. (1853107719) Examines the consequences of diminishing fish populations, and what impact that will have upon industries and nations. Read a related book at [pompeii.nap.edu/catalog/catalog.cfm?record_id=6032]. See an article and bibliography on water and fisheries at [www.gn.apc.org/eco/resguide/1_4ix.html]. See a related article at [www.emagazine.com/july-august_1996/0796feat1.html].

821. Wilson, Edward O., ed. *The Diversity of Life*, reissue ed. Cambridge: Harvard University, 424 pp., 1999 (originally published in 1992). (0393319407) How life came to evolve with such diverse results. The author says that we are now in the 6th great mass extinction: the human period. Reviewed at [www.igc.org/intheamazon/bkrev.htm]. See a bibliography of biodiversity at [management.canberra.edu.au/~gkb/habitatecos.html] and [www.uscusa.org/resources/biodiv.science.html].

822. Wilson, Edward O., and Perlman, Dan L. *Conserving Earth's Biodiversity* [CD-ROM computer program]. Washington, DC: Island Press, 1999.

(1559637730) Website links, photos, video-clips by Wilson, and essays. See a website about biodiversity at [kola.dcu.ie/~enfo/fs/fs10.htm].

823. Wolch, Jennifer, and Emel, Jody, eds. *Animal Geographies: Place, Politics, and Identity in the Nature-Culture Borderlands*. New York: Verso, 310 pp., 1998. (1859841376) Two professors of geography wrote this collection of essays: "consideration of the places where people and animals confront the realities of coexistence on an everyday basis, by way of case studies." "[A]nimals have been so indispensible to the structure of human affairs and so tied up with our visions of progress and the good life that we have been unable to (even try to) fully see them"; "how, by looking through geographical lenses, we may be able to bring animals into clearer focus."

824. Wolff, Pat, and St. John, Julie. *Waste, Fraud & Abuse in the U.S. Animal Damage Control Program*, 2nd ed. Wildlife Damage Review, 1999. Read the whole work online at [www.azstarnet.com/~wdr/wfa.html]. See also [www.fund.org/facts/adc.html] and [www.nets.com/fguardians/ adc.htm]. Read the Animal Damage Control Act at [ipl.unm.edu/cwl/fedbook/adca.html].

825. Yaffee, Steven Lewis. *Prohibitive Policy: Implementing the Federal Endangered Species Acts*. Cambridge: MIT, 240 pp., 1982. (0262240246) Read the table of contents and a brief review at [mitpress.mit.edu/book-home.tcl?isbn=0262240246]. Read an article in favor of the Endangered Species Act at [www.edf.org/EDF-Letter/1993/Mar/m_endspec.html].

Chapter 6

Animal Speculations

"What is truth?" asked Pontius Pilate of Jesus, according to the New Testament. What a tough question! This chapter is titled Animal Speculations, not because the resources in it are necessarily untrue. I am not of the belief that only scientifically provable things are true; I am a religious person. The fact is that many non-observable things are probably true, they are just not easy to prove. Thus, they are speculations.

This chapter contains nearly 100 references to books that explore the ideas of animal rights from a religious or "paranormal" perspective.

In general, animal rights supporters have viewed Christianity as hostile to their cause. This negative perception is not entirely unfounded, but neither is it wholly accurate.

The Christian religion had a significant impact on Western civilization. The weakening Roman empire made Christianity the state religion, enforced through the power of the emperors. The demise of Rome as a political power led to the rise of Rome as a religious and political power. The Roman Catholic church dominated the moral and social shape of Western culture until the fragmentation of sects in the Reformation, beginning around 1517.

While Christianity did espouse an improper view of "dominion," linking it to conquest and dictatorial powers over the animal kingdom, it also promoted some forms of ethics that included animal welfare. Constantine provided the world's first animal welfare laws from a political realm: (Judaism in the Torah did form some animal protection laws also), protecting horses from cruelty, in the fourth century A.D. Christians, though they might believe themselves to be dominant over the animals, do not generally practice cruelty intentionally. The best example of a Christian in close harmony with animals would be St. Francis of Assisi, a monk who lived during the Middle Ages. Not until the rejection of religions, in the seventeenth and eighteenth-century periods of Rationalism, did cruelty become permissible. It was decided (consciously or unconsciously) that ethics were by nature religious and unprovable, and therefore unnecessary to science. It was anti religion that led to the rise of vivisection and animal experimentation, and the idea that animals could not feel pain.

Nevertheless, Christianity has not supported animal-rights ideas very often. Thus, many animal-rights proponents have adopted Eastern relig-ions, pantheistic and Gaia-type religions, for their banners. One notable exception to this rule in modern times is that of Andrew Linzey, who has been actively promoting an animal-rights philosophy in conjunction with a form of Christian theology.

There is a growing movement in Judaism toward vegetarianism, and according to one source, the nation of Israel has the largest percentage of vegetarians in the world. Roberta Kalechofsky represents a primary source of such teachings from a Jewish point of view. While animal-rights activists condemn Judaism's practice of kosher slaughter, many believe it to be a rather painless manner of death for the animal. Special training is required to be a kosher butcher.

Aside from formal religion, there are also superstitions and myths to be considered. Many "New Age" religions view animals as angels, or reincarnated persons, or omens of good or ill news. Believers in the paranormal claim to have seen and spoken with deceased pet ghosts. Some also believe that they communicate with living animals through telepathy.

There are also a few miscellaneous items of speculation in this chapter, which could not be easily categorized in the other chapters. For example, do snakes really know or sense ahead of time that an earthquake is coming?

---

826. Anderson, Allen, and Anderson, Linda, eds. *Angel Animals: Exploring Our Spiritual Connection with Animals*. Plume, 256 pp., 1999. (0452280729) The authors founded an organization called Animal Angels "devoted to raising awareness about the spiritual relationship between people and animals." Anecdotal stories about the spiritual truths that animals teach us. Read some reviews and excerpts from the book at [www.angelanimals.com/books.html].

827. Andrews, Ted. *Animal Speak: The Spiritual and Magical Powers of Creatures Great and Small*. St. Paul, MN: Llewellyn, 383 pp., 1996. (0875420281) A shaman-style religious dictionary of animal, bird, and reptile symbolism. What animals mean in our dreams. See the table of contents at [www.newvision-psychic.com/bookshelf/animalspeak.html].

Read a brief review by the author at [www.totembooks.com/ani speak2.html].

828. Aylesworth, Thomas G. *Animal Superstitions.* New York: McGraw-Hill, 120 pp., 1981. (0070026580) See a bibliography of animal myths at [members.aol.com/karlshuker/3bibliography.html], and links to similar sites at [www.pibburns.com/cryptozo.htm].

829. Bakken, Peter W., et al. *Ecology, Justice, and Christian Faith: A Critical Guide to the Literature.* Bibliographies and Indexes in Religious Studies series, number 36. Westport, CT: Greenwood Press, 256 pp., 1995. (03132-90733) An annotated bibliography with more than 500 entries on environmental issues as addressed by Christianity in recent works. Use a similar bibliography online at [www. earthministry.org/anotebib.htm]. See a list of Christian organizations that teach environmentalism and ecology at [cesc.montreat.edu/ceo/index.html]. The Au Sable Institute is a well-known institution of this type, where students can take classes and receive degrees in ecology (with a Christian philosophy backdrop).

830. Barad, Judith A. *Aquinas on the Nature and Treatment of Animals.* Bethesda, MD: International Scholars, 195 pp., 1995. (1573090069) The author (a professor at Indiana State University) says that Aquinas held some early evolutionary concepts, and Barad seeks to explain the inconsistencies that appear between those writings and his ethical writings. Aquinas had a good deal to say about the similarities between man and animals, and also on the ethical treatment that animals should receive. Aquinas was probably the second most influential Christian writer in history, after St. Augustine. See a related article at [custance.org/evol/ 5intro.html].

831. Bardens, Dennis. *Psychic Animals: A Fascinating Investigation of Paranormal Behavior.* New York: Barnes and Noble, 203 pp., 1996 (originally published in 1987). (080500730x) Stories of psychic animals, including cases of extrasensory perception (ESP) and evidences of altruism. See a site with discussion of psychic animals at [www.sylvia.com/SyteBeyond/psychicanimals1.htm].

832. Berry, Rynn. *Food for the Gods: Vegetarianism and the World's Religions.* New York: Pythagorean, 392 pp., 1998. (0962616923) Reviewed at [www.ivu.org/books/reviews/food-for-the-gods.html]. Essays and interviews with Hindus, Christians, Taoists, Muslims, Jews, and others who practice vegetarianism. Religion has been one of the staunchest supporters of human carnivorism, promoting and encouraging the eating of meat. In Christianity, this is often described as having "dominion" over the

creation. On the other hand, a plain reading of Genesis would indicate that humankind was vegetarian and was only granted permission to eat meat after the Great Flood. Still, there has always been at least a minority of religious people who have maintained vegetarian diets for various reasons.

833. Borowski, Oded. *Every Living Thing: Daily Use of Animals in Ancient Israel*. Walnut Creek, CA: Altamira, 296 pp., 1998. (0761989188) This information is useful to Jewish, Christian, and Islamic researchers who would study the use of animals by their spiritual ancestors. Read a review at [www.wisc.edu/larch/sas/9901k.htm].

834. Bratton, Susan Power. *Christianity, Wilderness, and Wildlife: The Original Desert Solitaire*. Scranton, PA: University of Scranton, 352 pp., 1993. (0940866145) "[T]he long history of Christian wilderness spirituality and beneficial human interactions with wild nature." (dust jacket flap) She looks at biblical and early church history. The author has a Ph.D. in ecology from Cornell University. Read an article by the author at [www.vts.edu/vsjournal/December1997/desertsolitaire.htm]. Reviewed at [asa.calvin.edu/ASA/book_reviews/9-94.htm].

835. Buckner, E. D. *The Immortality of Animals*, reprint ed. Mellen Animal Rights Library #H6. Lewiston, NY: Mellen, 291 pp., 1998 (originally published in 1903). (0773487263) An excellent attempt to prove that animals are given an afterlife by God. See a very brief summary (and order a spiral-bound version of the book) at [creatures.com/Immor tality.html]. Read related articles at [www.geocities.com/Athens/ Oracle/1783/an.htm] and [www.dailyegyptian.com/spring97/042897/ souls.html]. This is the best book I have read on the question of animal immortality.

836. Buddemeyer-Porter, Mary. *Will I See Fido in Heaven? Scripturally Revealing God's Eternal Plan for His Lesser Creatures*. Shippensburg, PA: Companion Press, 106 pp., 1995. (1560435534) "Many Scriptures give me peace concerning the future of my pets in Heaven." (p. 93) Read excerpts and see some links at [www.creatures.com/SiteMap. html].

837. Carr-Gomm, Philip and Stephanie. *The Druid Animal Oracle: Working with the Sacred Animals of the Druid Tradition*. New York: Simon & Schuster, 184 pp., 1994. (0671503006) How to use 33 sacred Druid animals as oracles to change your life. The book includes a set of animal cards, which are to be used in a Tarot-style oracle reading. See a long excerpt at [druidry.org/obod/text/lewes/archive/pcg-dao.html].

838. Corless, Roger. *The Concern for Animals in Buddhism*. Silver Spring, MD: International Network for Religion and Animals, 1991? See articles about

animals in Buddhism at [online.sfsu.edu/~rone/Buddhism%20and%20 Animal%Rights.htm] and [www.buddhism.ndirect.co.uk/fstory4.htm].

839. Fate Magazine. *Psychic Pets & Spirit Animals*. St. Paul, MN: Llewellyn, 272 pp., 1996. (1567182992) 36 stories taken from *Fate Magazine*. Real-life experiences with common and uncommon creatures on psychic communication, life after death, ghosts, omens, and so on. Find a test questionnaire for your pets at [orion.csuchico.edu/Pages/vol39issue05/ d.psychicpets.html].

840. Fox, Michael W. *The Boundless Circle: Caring for Creatures and Creation*. Wheaton, IL: Quest, 300 pp., 1996. (0835607259) The author brings out the religious dimension of animal rights ideas in a scholarly way. He hopes that modern religions, especially Christianity, will adopt "panentheism" (similar to pantheism, the belief that God is in everything) so that our attitudes toward nature and animals will change. We must recognize our "kinship" to animals. See an article about the rise of global green religion at [www.geocities.com/Capitol Hill/Senate/4904/ggreligion. htm].

841. Fox, Michael W. *Lessons from Nature, Dr. Fox's Fables*. Washington, DC: Acropolis, 157 pp., 1980. (0874912911) 23 fables in which animals talk about how they feel and live.

842. Gellatley, Juliet. *Going for the Kill: Viva! Report on Religious (Ritual) Slaughter*. Brighton: Viva!, 1998. Read the lengthy report at [www.viva. org.uk/Viva!%20Campaigns/Slaughter/GoingfortheKill.htm].

843. Goldstein, Martin. *The Nature of Animal Healing: The Path to your Pet's Health, Happiness and Longevity*. New York: Knopf, 352 pp., 1999. (0679455000) The author has a degree from Cornell University in Veterinary Medicine, and 25 years of practice. In this book he promotes holistic medicine for pets: diet, acupuncture, homeopathy, and so on. Read an interview with the author at [www.randomhouse.com/knopf/aak/qna/ goldstein.html].

844. Hall, Manly P. *The Inner Lives of Minerals, Plants, and Animals*, reprint ed. Los Angeles: Philosophical Research Co., 31 pp., 1999 (originally published in 1973). (089314227) The author founded the Philosophical Research Society in 1934. This sounds similar to the traditional Native American view of the Great Spirit indwelling all material objects, including trees.

845.Hamilton, Joseph. *Animal Futurity*, reprint ed. Mellen Animal Rights Library #H5. Lewiston: Mellen, 1997 (originally published in 1877). (0773487247) A just God must recompense animals for their undeserved sufferings. This is not a radical idea for Christian theology; C. S. Lewis discusses it at length in two books that I have listed below.

846.Hillman, James. *Dream Animals*. San Francisco: Chronicle, 90 pp., 1997. What our dreams show about our views on animals. Reviewed at [www.montelis.com/satya/backissues/march98/dream.html].

847.Hume, C. W. *The Status of Animals in the Christian Religion*, reprint ed. London: Universities Federation for Animal Welfare (UFAW), 109 pp., 1980 (originally published in 1957). "The author notes the scanty importance attached to the welfare of animals by many theologians and sets out to enquire how this habit of mind came about." (dust jacket flap) The author founded the UFAW. See the UFAW webpage at [www.ufaw3.dircon. co.uk/].

848.Hunt, Robert. *Is My Dog in Heaven? A Biblical Answer*. New York: Vantage, 49 pp., 1995. (0533114691) By a Presbyterian minister. See a bibliography at Dogheaven Bookstore [www.dogheaven.com/BOOK STORE.HTM].

849.Hyland, J. R. *The Slaughter of Terrified Beasts: A Biblical Basis for the Humane Treatment of Animals*. Sarasota, FL: Viatoris, 86 pp., 1988. (0945703007) "The Bible calls upon human beings to stop their violence and their abuse of each other, and of all other creatures." (back cover) See an article at [www.leaderu.com/orgs/probe/docs/ anim-rts.html].

850.Isaacs, Ronald H. *Animals in Jewish Thought and Tradition*. Northvale: Jason Aronson, 1999. (0765799766) Not yet published. See a related website with articles at [www.earth.org.hk/jewprop.html] and an article at [www.mnsinc.com/dsvtx/ creatures.html].

851.Kadletz, Edward. *Animal Sacrifice in Greek and Roman Religion*. Seattle: University of Washington, 1976. See the full annotation in chapter 3 (on Fatal Uses of Animals).

852.Kalechofsky, Roberta, ed. *Judaism and Animal Rights: Classical and Contemporary Responses*, reprint ed. Marblehead, MA: Micah Publications, 356 pp., 1994 (originally published in 1992). (0916288358) 41 articles, ancient and modern, on Jewish traditions of concern for animals. "[T]saar balei chayim — you may not cause sorrow to living creatures." Divided into sections: how we think, how we eat, and how we are now.

Read the table of contents at [www2.upperaccess.com/ upperaccess/1023-toc.htm]. See a resume and biography of the author at [www. micahbooks.com/speaker.html].

853. Kalechofsky, Roberta, ed. *Rabbis and Vegetarianism: An Evolving Tradition*. Marblehead, MA: Micah Publications, 104 pp., 1995. (09162-88358) 17 essays by Rabbis. See links and articles at [www.all-creatures. org/articles/jvshabbat.html], [www.rasheit.org/NY-REBBES/rebschwartz. html] and [www.vegsource.org/biospirituality/xtianity. html].

854. Kalechofsky, Roberta, ed. *Vegetarian Judaism: A Guide for Everyone*. Marblehead, MA: Micah Publications, 246 pp., 1998. (0916289455) Uses moral and scriptural principles of Judaism to promote vegetarianism. The author says that vegetarianism guards your health, does not cause pain to animals, uses Earth's resources more prudently, helps the poor, and strengthens community life: all principles of Judaism. Reviewed at [www.micahbooks.com/jewishvegbooks.html] and [www. rasheit.org/VY-LIBRARY/libschwartz20.html].

855. Kapleau, Roshi Philip. *To Cherish All Life: A Buddhist Case for Becoming a Vegetarian*. New York: Harper & Row, 104 pp., 1982. (094030600x) The author had 13 years of formal training in Japan in Zen Buddhism. Read an article at [www.buddhanet.net/fdd21.htm] and a link to a long critique at [www.buddhanet.net/ftp11.htm].

856. Kolb, Janice Gray. *Compassion for All Creatures: An Inspirational Guide for Healing the Ostrich Syndrome*. Nevada City: Blue Dolphin, 264 pp., 1997. (1577330080) "[A]himsa or non-violence . . . is firmly entrenched in the compassion that our Lord placed within my heart for humans and animals alike." She says that religions demand compassion, which should be extended to animals. Read about ahimsa at [www.ettl.co.at/uc/ws/ theme034] and a brief publisher review at [www.vpg.net/ostrich.htm].

857. Kowalski, Gary. *The Souls of Animals*, 2nd revised ed. Walpole: Stillpoint, 160 pp., 1999 (originally published in 1991). (0913299847) Universalist Unitarian minister at Harvard; "animals are not inanimate objects devoid of feeling and intellect but thinking, sentient individuals with a spiritual life." One chapter discusses the artistic interests of elephants, referring often to the book by David Gucwa and James Ehmann, *To Whom It May Concern*, found in chapter 3 of this bibliography. Kowalski says that playfulness, love, awareness of death, and altruism are all spiritual qualities in animals. Read a related article at [www.blavatsky.net/ blavatsky/arts/HaveAnimalsSouls.htm].

858.Kurz, Gary. *Cold Noses at the Pearly Gates: A Book of Hope.* Kearne: Gary Kurz, 116 pp., 1997. (0966611705) This retired Coast Guard officer uses strict Biblical interpretations to comfort sad Christians whose pets have died. He purposely avoids speculations, but concludes that animals go to Heaven. He is critical of other Christian writers and their opinions on this subject. To see many short reader reviews of the book (now more than two dozen) search "Kurz and Cold Noses" on [www.amazon.com].

859.Laland, Stephanie. *Animal Angels: Amazing Acts of Love and Compassion.* Berkeley, CA: Conari, 219 pp., 1998. (1573241423) Very short anecdotes that may suggest altruism (love) in animals. The author says that her goal is to show the spiritual interconnectedness of all living organisms. Find the publisher review and author biography at [www.conari.com/BookPages/BK_AnimalAngels.htm]. Reviewed at [www.montelis.com/satya/backissues/sep98/mercy.html].

860.Lewis, C. S. *God in the Dock: Essays on Theology and Ethics*, reprint ed. Grand Rapids, MI: Eerdmans, 346 pp., 1994 (originally published in 1970). (0802808689) Includes an extensive discussion with C.M. Joad on animal pain and animal immortality. For an excellent summation read an article by Andrew Linzey called "C.S. Lewis's Theology of Animals" in *Anglican Theological Review*, vol. LXXX, no. 1, Winter, 1998. Read an extensive article about the bioethics of C.S. Lewis at [people.ne.mediaone.net/haasj/ethics/PSCFLeBar.html].

861.Lewis, C. S. *The Problem of Pain* (1962), reprint ed. New York: Collier, 160 pp., 1986 (originally published in 1962). (002086502) A dramatic book written shortly after the death of his wife. In contemplating the nature and reasons why God allows painful events in human lives, he also asks questions about the pain of animal lives. Does God recompense animals for the suffering humans cause them?

862.Linzey, Andrew. *Animal Gospel: Christian Faith as Though Animals Mattered.* London: Hodder & Stoughton, 184 pp., 1999. (0340621508) The book "makes a strong and moving case for a radically reformed faith which allows the Gospel of the crucified Christ to interpret the world of innocent suffering."(back cover) The author also has a related audio cassette sermon available called "Not a Sparrow Falls: Gospel Truths about Animals," available from Anchor Recordings Ltd., 72, The Street, Kennington, Ashford, Kent, TN24 9HS U.K.

863.Linzey, Andrew. *Animal Rights: A Christian Assessment of Man's Treatment of Animals.* London: SCM Press, 120 pp., 1976. The author says that vegetarianism is a must in Christianity. He critiques vivisection and discusses the views of Thomas Aquinas and Albert Schweitzer. See a good bibliography at [www.cep.unt.edu/ecotheo.html]. Read an excerpt at [www.ivu.org/people/writers/linzey.html].

864. Linzey, Andrew. *Animal Rites: Liturgies of Animal Care.* Harrisburg, PA: Trinity Press International, 144 pp., 1999. (0334027608) "This book provides fourteen new liturgies that are animal-friendly and animal-inclusive. They include services of celebration for animal companionship, services for animal welfare, healing liturgies, new eucharistic prayers 'for the whole creation,' and animal burial services." Briefly reviewed at [ely.anglican.org/parishes/camgsm/Majestas/1999/April.html#animal_rites_by_andrew_linzey].

865. Linzey, Andrew. *Animals and Christianity: A Book of Readings.* New York: Crossroad, 210 pp., 1988. (0824509021) 50 readings intended for college classes. Read an article by the author at [www.ivu.org/news/1-96/linzey.html].

866. Linzey, Andrew. *Animal Theology.* Chicago: University of Illinois, 214 pp., 1995. (0252021703) The author rejects humanocentrism and he presents humans as servants, protecting the weak. Linzey calls hunting "the anti-gospel of predation." He has written more than a dozen books on Christian ethics and theology, especially in relation to animals. Chapters include: "Humans as the Servant Species," "Animal Experiments as Un-Godly Sacrifices," "Vegetarianism and the Biblical Ideal," and "Genetic Engineering as Animal Slavery." Read a publisher review at [www.press.uillinois.edu/s95/linzey.html].

867. Linzey, Andrew. *Christianity and the Rights of Animals.* New York: Crossroad, 191 pp., 1987. The author interprets dominion and covenant in pro animal ways and says that God has a right to demand respect for animals. See links on the moral status of animals at [www.acusd.edu/ethics/animal.html]. See an extensive article on religion and animal rights at [www.api4animals.org/Publications/AnimalIssues/1996-Spring/ReligionGiefer.htm].

868. Linzey, Andrew, and Cohn-Sherbok, Dan. *After Noah: Animals and the Liberation of Theology.* New York: Cassell, 128 pp., 1997. (0264674502) "The Jewish and Christian traditions are often blamed for apparently justifying the abuse of animals. But . . . there are also considerable resources within both traditions to support an enlightened and ethical view of animals." Read a short article by Linzey at [arrs.envirolink.org/ar-voices/Christian.html]. Cohn-Sherbok teaches Judaism at the University of Wales in Lampeter.

869.Linzey, Andrew, and Yamamoto, Dorothy, ed. *Animals on the Agenda: Questions about Animals for Theology and Ethics*. Chicago: University of Illinois, 320 pp., 1998. (0252067614) Read an excerpt at [www.press. uillinois.edu/f98/excerpts/linzey/intro.HTML]. "This encyclopaedic volume is the most comprehensive collection of original studies on animals and theology ever published. With contributors from both sides of the Atlantic."

870.Lodrick, Deryck O. *Sacred Cows, Sacred Places: Origins and Survivals of Animal Homes in India*. Berkeley: University of California, 307 pp., 1981. (0520041097) The first systematic attempt to describe the Hindu institution of Goshala in a cultural framework; homes for old or infirm cattle. Many in India believe that cows and other creatures are reincarnated forms of ancestors, and thus they are not killed. Read an article about sacred cows in India at [www.hinduism.co.za/cowsare.htm].

871.Majpuria, Trilok Chandra. *Sacred Animals of Nepal and India: With Reference to Gods and Goddesses of Hinduism and Buddhism*. Lashkar: M. Devi, 1991. Though the Hinduism of old believed in many gods; more modern Hindus tend to view the many idols and sacred animals as different representations of the same one god. Read a tongue-in-cheek (but factual) article about the situation of rats in India at [home.plex.nl/ ~omniron/wrindiatemple.html]. See a large compilation of Hindu and Buddhist texts regarding animals at [www.tparents.org/Library/ Unification/Books/World-S/WS-05-02.htm].

872.Manes, Christopher. *Other Creations: Rediscovering the Spirituality of Animals*. Garden City: Doubleday, 288 pp., 1997. (0385483651) How animals affect human spirituality and the human psyche. Read an interview with the author at [www.bestfriends.org/gc/manes.htm]. See a relevant article at [augustachronicle.com/stories/050397/fea_pets. html].

873.Masri, Al-Hafiz B. A. *Animals in Islam*. Petersfield, England: Athene Trust, 1989. The author also created a 27-minute video called "Creatures of God." Use a wonderful site on the subject of animals in Islam at [www.geocities.com/Athens/Academy/7368/an1.htm]. Read an interview with a Moslem theologian on the subject at [animals.co.za/orgs/ animalvoice/98juloct/theology.html].

874.McDaniel, Jay Byrd, and Cobb, John B., Jr. *Of God and Pelicans: A Theology of Reverence for Life*. Philadelphia: Westminster, 168 pp., 1989. (0664250769) Read a Christian sermon which offers both pro and con views of animal rights at [home.msen.com/~firstuu/min_msg/sermons/

anmlrgts.txt]. Use Cobb's online ecotheology bibliography at [www.cep.unt.edu/ecotheo.html].

875.Murti, Vasu. *They Shall Not Hurt or Destroy: Moral and Theological Objections to the Human Exploitation of Nonhuman Animals.* Oakland, CA: Vasu Murti, 199 pp., 1995. This is an excellent punchbound book, looking historically at theological and religious views on animal treatment. There are a lot of surprising quotations and ideas to be found here, mainly from the Jewish, Christian, and Islamic traditions. Order a copy direct from the author at 30 Villanova Lane, Oakland, CA, 94611 ($20 includes the shipping). Read an article by the author at [www.fnsa.org/v1n4/murti1.html]. See a similar article at [arrs.envirolink.org/ar-voices/hunting_scripture.html].

876.Myers, Arthur. *Communication with Animals: The Spiritual Connection Between People and Nature.* Neptune City: NTC, 256 pp., 1993. (08092-31492) Case studies of nonverbal communication by animals with humans; interviews with 50 animal communication experts. The author says that animals and people are telepathic. See a review and excerpts at [www.globalpsychics.com/lp/animalstalk/ arthowto.htm].

877.Naganathan, G. *Animal Welfare and Nature: Hindu Scriptural Perspectives.* Washington, DC: Center for Respect of Life and Environment, 25 pp., 1989. Related articles at [www.ivu.org/articles/net/hinduism.html] and [www.ivu.org/articles/net/hindus.html].

878.Nitsch, Twylah. *Creature Teachers: A Guide to the Spirit Animals of the Native American Tradition.* New York: Continuum, 104 pp., 1997. (0826410235) See a brief biography and review at [www.continuum-books.com/womsd7.htm].

879.Page, Ruth. *The Animal Kingdom and the Kingdom of God.* An Occasional Paper from the Centre for Theology and Public Issues. Edinburgh: Edinburgh University, 42 pp., 1991. (1870126173)

880.Palmer, Bernard, and Jones, M. L. *What Are They Trying to Do to Us? The Truth About the Animal Rights Movement & the New Age.* Alpharetta, GA: John Honea, 1994. (1882270118) I have also seen this listed as being published by Old Rugged Cross Press in 1994.

881.Pasten, Laura. *The Tarantula Whisperer: A Celebrity Vet Shares Her Secrets in Communication with Animals.* Berkeley: Conari Press, 200 pp., 1999. (1573241598) A fun and interesting book of veterinarian stories. The author takes a very mystical view of communication, saying that we all can use telepathy to train and talk to our animals. See the author's page at [www.drpasten.com/index.html] and a similar article at [www.animal kinship.com/WithinArticle.htm].

882.Pinches, Charles, and McDaniel, Jay B. *Good News for Animals? Christian Approaches to Animal Well-Being*. Maryknoll: Orbis, 258 pp., 1993. (088-3448599)

883.Primatt, Humphrey. *A Dissertation on the Duty of Mercy and the Sin of Cruelty*, reprint ed. Mellen Animal Rights Library #H3. Lewiston, NY: Mellen, 326 pp., 1996 (originally published in 1776). (07734-87182) "Pain is pain, whether it be inflicted on man or on beast. . . . a man can have no natural right to abuse and torment a beast." The first major theological defense of animals. The author says that all sentient beings deserve mercy and justice.

884.Quaker Concern for Animal Welfare. *Regarding Animals: Some Quakers Consider Our Treatment of Animals in the Modern Age*. Braintree: Quaker Concern for Animal Welfare, 1985. Read an article by a vegetarian Quaker at [www.ivu.org/news/3-98/meat.html].

885.Regan, Tom ed. *Animal Sacrifices: Religious Perspectives on the Use of Animals in Science*. Ethics and Action series. Philadelphia: Temple University, 270 pp., 1986. (0877224110) "[T]his book presents the teachings of the major religions of the world concerning animals." (back cover) Included are passages from Christianity, Judaism, Islam, Hinduism, Jainism, Buddhism, and Confucianism. See a related site at [www. simr.dircon.co.uk/religion.html].

886.Regan, Tom. *The Struggle for Animal Rights*. Clarks Summit, PA: International Society for Animal Rights, 197 pp., 1987. (0960263217) The author speaks about the role of religion in shaping culture and improving animal lives and also seeks to find the major causes of animal exploitation in our world.

887.Regenstein, Lewis G. *Replenish the Earth: A History of Organized Religion's Treatment of Animals and Nature*. New York: Crossroad, 304 pp., 1991. (0824510755) The author was nominated for a Pulitzer Prize for his work *America the Poisoned*. This is the best summary of the religious history of animal-treatment that I have read. Includes looks at Christianity, Hinduism, Islam, Judaism, and Buddhism. Reviewed at [www.rasheit.org/ VY-LIBRARY/libschwartz21.html# anchor2710919].

888. Robson, Frank. *Pictures in the Dolphin Mind.* New York: Sheridan House, 135 pp., 1988. (0911378782) This man worked at Marineland with captive dolphins for many years. This book talks about dolphin telepathy. Read excerpts at [www.dolphininstitute.org/isc/text/ e_telpth.htm]. See related articles on captive dolphins at [www.pbs.org/wgbh/pages/frontline/ shows/whales/man/comm.html] and [www.dolphininstitute.org/isc/text/ e_capt.htm].

889. Salgia, T. J. *Jainism, Non-Violence and Vegetarianism.* Salgia, 1987. Jainism is a non violent religion disallowing any intentional causing of death. Even insects are exempt from being killed by Jains. Read an article at [www.ivu.org/articles/net/ jainism.html], and find several links at [www.cs.colostate.edu/~malaiya/jainhlinks.html].

890. Sargent, Tony. *Animal Rights and Wrongs: A Biblical Perspective.* London: Hodder, 264 pp., 1996. (0340669136) The author touches on many subjects, including pets, vegetarianism, animal pain, and animal abuse. Reviewed at [dspace.dial.pipex.com/town/plaza/jl51/archive/ 07f02.htm]. See articles about the "image of God" at [www.capo.org/ jbem/walker-w.htm] and [. . . walker-s.htm].

891. Saunders, Nicholas J. *Animal Spirits: The Shared World of Sacrifice, Ritual and Myth; Animal Souls and Symbols.* Living Wisdom series. Boston: Little, Brown, 184 pp., 1995. (0316903051) The book appears to be a scholarly presentation, not a mystical one, of the role of animals as symbols in human civilizations.

892. Schaeffer, Francis. *Pollution and the Death of Man: The Christian View of Ecology*, reprint ed. Wheaton: Crossway, 125 pp., 1992 (originally published in 1970). (0891076867) The author predicted that Christian apathy to the environment and animals would lead future generations to accept Eastern religions. See author information at [francis.schaeffer.net/]. Read an article on Christian environmentalism at [leaderu.com/orgs/probe/docs/ ecology.html]. Schaeffer was correct: the Christian church's denouncing of environmental and animal-rights issues has sent most such people into eastern philosophies and religions.

893. Schochet, Elija Judah. *Animal Life in Jewish Tradition: Attitudes and Relationships.* New York: KTAV, 379 pp., 1984. (0881250198) Read an article about ritual kosher slaughter at [lamar.colostate.edu/~grandin/ ritual/kosher.slaugh.html].

894. Schoen, Allen M., and Proctor, Pam. *Love, Miracles, and Animal Healing: A Heartwarming Look at the Spiritual Bond Between Animals and*

Humans. New York: Simon & Schuster, 238 pp., 1995. (0684802074) A vet's journey from physical medicine to spiritual understanding; he offers folk remedies and practices holistic medicine. See the author's site with many of his articles at [www.gcci.org/ciah/ciah_schoen.html]. Read a bibliography of homeopathic pet care books at [www.morrills.com/ pet_books.htm].

895. Schul, Bill D. *Animal Immortality: Pets and Their Afterlife*. New York: Carrol & Graf, 221 pp., 1990. (0881845418) "Dr. Schul here presents remarkable and fully documented case studies of domestic and other animals whose spirits have been known to return to communicate with their former masters or haunt their past environments." (back cover) Reviewed at [www.physics.helsinki.fi/whale/intersp/pages/ book1.htm].

896. Schul, Bill D. *The Psychic Power of Animals*. Greenwich, CT: Fawcett, 223 pp., 1977. (0449137244) "He [the animal] is more sensitive to psychic phenomenon, the presence of apparitions, disembodied spirits; he can monitor happenings hundreds of miles away." (p. 12) Related articles and links at [www.cyberark.com/animal/telepath.htm].

897. Sheldon, Joseph Kenneth. *Rediscovery of Creation: A Bibliographical Study of the Church's Response to the Environmental Crisis*. American Theological Library Association (ATLA) Bibliography series #29. Metuchen, NJ: Scarecrow, 282 pp., 1992. (0810825392) A briefly annotated bibliography of books and magazine articles which provide a historical overview of the contribution of Christianity to environmental (and tangentially, animal rights) issues. Reviewed at [asa.calvin.edu/ ASA/book_reviews/9-93.htm].

898. Siddiqui, Muhammad Iqbal. *The Ritual of Animal Sacrifice in Islam*. Lahore: Kazi Publications, 47 pp., 1993. (0935782362) Read a brief article on the subject at [www.arabicbible.com/islam/sacrif.htm] and a related article at [www.rediff.com/news/1998/apr/16varsha.htm].

899. Smith, Penelope. *Animal Talk: Interspecies Telepathic Communication*, 2nd ed. Hillsboro: Beyond Words, 150 pp., 1999. (158270001x) See a list of related courses taught by the author at [207.155.27.19/cyberark/ animal/courses.htm]. See her telepathy site (chiefly with dolphins) at [www.divinedolphin.com/].

900. Smith, Scott S. *The Soul of Your Pet: Evidence for the Survival of Animals in the Afterlife*. Washington, DC: Holmes Publishing Group, 120 pp., 1998. (1558184023) Read a brief review by the author at [www. arktrust.org/animals/books/petsouls.htm].

901. Sobosan, Jeffrey G. *Bless the Beasts: A Spirituality of Animal Care*. New York: Crossroad, 144 pp., 1991. (0824511352) "Christianity must abandon Greek metaphysics to restore due reverence to animals." (The author is a Catholic theology professor at the University of Portland, Oregon. He is a process theologian, and strongly opposes "speciesism". See a negative and brief review of this and another Sobosan book at [www.leaderu.com/ftissues/ft9706/briefly.html].

902. Sorrell, Roger D. *St. Francis of Assisi and Nature*. New York: Oxford Univ., 204 pp., 1988. (0819908827) A fascinating explanation of the beliefs, life, and influence of St. Francis of Assisi, a medieval Christian monk who brought a love of nature to the church. Read about Francis at [www.stolaf.edu/president/francis3.htm]. See a short article on St. Francis at [www.imma.org/misanthrope.html].

903. Stefanatos, Joanne. *Animals and Man: A State of Blessedness*. MN: Light and Life, 344 pp., 1992. (0937032905) 52 true stories in history of holy men and women who lived with wild animals. The author is a pioneer of holistic veterinary medicine.

904. Steiger, Sherry Hansen, and Steiger, Brad. *Mysteries of Animal Intelligence*. New York: Tor, 194 pp., 1995. (0812551915) Anecdotal evidence of animal intelligence, which leans toward paranormal explanations. The authors have written a couple of sequels to this book. Read an interview with the authors at [www.insight2000.com/steiger.html]. See a lengthy article in Adobe Acrobat format at [www.uwm.edu/Dept/Honors/animalthinking/esp.pdf].

905. Sutton, John. *Psychic Pet: Supernatural True Stories of Paranormal Animals*. Hillsboro: Beyond Words, 124 pp., 1998. (188522379x) Stories of cats, dogs, and horses capable of telepathy, teleporting, and returning to earth as ghosts. In the book is included a questionnaire for pet owners to use to determine if their own animals are psychic. Read an excerpt at [www.invink.com/x675.htm]. See an extensive site with many links at [www.argonet.co.uk/users/lyndale/lotcaf/yiffle/spiritual/animal/anmlcmmcte/index.htm].

906. Svedbeck, Frances. *Do Animals Go to Heaven?* Manchester: Eden, 127 pp., 1994. "It is my earnest desire that this book will relieve the nagging fear that many animal lovers have, concerning their pet's eternal destiny." See more on this self published book at [www.creatures.com]. Read a similar online book at [www.all-creatures.org/bookallcreatures.html].

907. Thompson, James. *Cast Out of the Ark.* Holywell, England: Christians Against All Animal Abuse, 1989. I have not been able to find more details on this work. See a webpage with links about animals in the religions of the world at [www.argonet.co.uk/users/lyndale/lotcaf/yiffle/spiritual/animal/index.htm].

908. Toperoff, Shlomo Pesach. *The Animal Kingdom in Jewish Thought.* Northvale: J. Aronson, 269 pp., 1995. (1568214391) A study of 65 animal species listed in the Bible, analyzed in alphabetical order. The author (a Rabbi) draws information from rabbinical and biblical literature. Includes introductory chapters on animal welfare, animals in art, and veterinary surgery.

909. Tributsch, Helmut. *When the Snakes Awake: Animals and Earthquake Prediction.* Cambridge: MIT, 248 pp., 1984 (originally published in 1982). (0262700255) Read a related article at [www.levity.com/mavericks/quake.htm]. This is a folk legend that should not be too easily dismissed. I have lived in California's earthquake-prone regions and have heard animals carrying on loudly before earthquakes have struck. Do they hear some ultrasonic evidence of the coming quake? I have no idea; but they do seem to get nervous about something.

910. Von Kreisler, Kristin. *The Compassion of Animals: True Stories of Animal Courage and Kindness.* Rocklin: Prima, 257 pp., 1997. (076-1509909) Brief reviews at [www.pighealth.com/MEDIA/P/BOOKS/BKDETAIL/KREISLER.HTM].

911. Waddell, Helen, translator. *Beasts and Saints,* reprint ed. Grand Rapids: Eerdmans, 132 pp., 1996 (originally published in 1934). (0802842232) "[S]tories of animals and saints, taken from the Desert Fathers and the Celtic saints . . . in the early days of the Christian heritage and the holistic spirituality that belongs to that world." Read an excerpt at [www.wordtrade.com/Religion/saints.htm].

912. Webb, Stephen H. *On God and Dogs: A Christian Theology of Compassion for Animals.* New York: Oxford University, 222 pp., 1998. (019511650x) "[A]gainst the more extreme animal liberationists — defends the intermingling of the human and animal worlds." This professor of religion from Wabash College says that Christian grace should be applied to our dealings with animals. Reviewed at [www1.christianity.net/bc/8B5/8B5006.html] and reviewed by the author at [www.montelis.com/satya/backissues/sep98/dogs. html].

913. Wiebers, Andrea Gillan, and Wiebers, David O. *Souls Like Ourselves: Inspired Thoughts for Personal and Planetary Advancement.* Rochester, MN: Sojourn Press, 202 pp., 2000. (0967097908) This is a collection of brief quotations on the subject of animals, chosen in a devotional or inspirational manner. The quotations chosen are marvelous. The only problem from a scholarship angle is that there are only short bibliographic references (only author and title), meaning that finding the original source would be difficult. Read an article by David Wiebers at [www.theosophy-nw.org/theosnw/issues/an-wieb2. htm].

914. Wrighton, Basil. *Reason, Religion and the Animals.* London: Catholic Study Circle for Animal Welfare, 1989. Order directly from 39 Onslow Gardens, South Woodford, London E18 1ND, United Kingdom.

915. Wylder, Joseph Edward. *Psychic Pets: The Secret Life of Animals.* New York: Bonanza, 161 pp., 1989. (0883730804) How animals understand us using extrasensory perception (ESP). See pages of psychics who claim to understand your pet's telepathic signals at [animalstalk.com/] and [animaltalk.net/animal%20communicators. htm]. See a brief article at [mkb-psychic.com/articles.htm#ANIM ALS].

916. Young, Richard Allen, and Adams, Carol J. *Is God a Vegetarian? Christianity, Vegetarianism, and Animal Rights.* Chicago: Open Court, 187 pp., 1998. (0812693930) A nicely written group of alternative perspectives and interpretations on the Christian/biblical view of meat-eating. Generally he promotes vegetarianism by reinterpreting common Bible passages, as opposed to the simple dismissal of all "disagreeable" passages that many vegetarians have practiced. See related articles at [www.ivu.org/news/95-96/religion.html], [members.aol.com/feloflife/feloflife.html], and [www. peta-online.org/jesus/ scholar.html].

Index

About the Author

JOHN M. KISTLER is Collection Development Librarian at West Virginia State College.